CLASSIC GEOLOGY IN EUROPE 4

Canary Islands

D1145762

Canary Islands

Juan Carlos Carracedo
Consejo Superior de Investigaciones Científicas

Simon Day
University College London

TERRA

© Juan Carlos Carracedo & Simon Day 2002

First published in 2002 by Terra Publishing

Terra Publishing
PO Box 315, Harpenden, Hertfordshire AL5 2ZD, England
Telephone: +44 (0)1582 762413
Fax: +44 (0)870 055 8105
Website: http://www.terrapublishing.co.uk
E-mail: publishing@rjpc.demon.co.uk

British Library Cataloguing-in-Publication Data
A CIP record for this book is available from the British Library

Library of Congress Cataloging-in-Publication Data are available

Typeset in Palatino and Helvetica
Printed and bound by Biddles Ltd, Guildford and King's Lynn, England

Contents

Preface

The Canary Islands are rarely thought of as Europe's highest volcanoes, perhaps because so much of their bulk is concealed beneath the surface of the ocean, but in fact they include some of the largest and most active volcanoes on the planet. Add to that the wide variety of volcanic rocks and structures that are present, the exceptionally good exposure and preservation of the rocks, and the pleasant climate, and it is easy to see why they are among the best places in the world to study volcanoes and volcanic processes.

Chapter 1 gives a general introduction to the geology of the islands and the mechanisms by which such oceanic island volcanoes form. Chapter 2 discusses the logistics of visiting the Canaries; it also contains guidelines for working safely and enjoyably in the islands, with special attention paid to the local climatic conditions. The remaining chapters each deal with one of the seven main islands of the Canaries, from the eroded roots of ancient volcanoes exposed in Fuerteventura, the oldest island, to the historical lava flows of the youngest islands, La Palma and El Hierro. Each island exemplifies a particular aspect of the evolution of oceanic-island volcanoes, from the initial growth of seamounts below sea level through the highly active shield stage of volcanism, to the final infrequent eruptions that punctuate the long erosion of the older islands. Each chapter brings out the implications of the geology of each island for the processes involved, both in the Canaries and in other oceanic-island groups. The Glossary and Further Reading follow at the end of the book. To maintain an informal style, relevant sources have not been referenced as they would be in a specialist journal or textbook; instead, key papers appear in the Further Reading. We trust that our friends and colleagues will indulge us in taking this mild professional liberty.

Juan Carlos Carracedo
Simon Day
July 2002

Acknowledgements

We would like to thank the many colleagues and friends who have shared the study of the Canary Islands with us, in particular Hervé Guillou, Francisco Pérez Torrado, Raphael Paris, Eduardo Rodríguez Badiola, Bruce Nelson, Hubert Staudigel, Giray Ablay, Martin Gee. Last, but by no means least, special mention must be made of the invaluable assistance provided by the publishers, and of their exceptional patience.

Chapter 1

Geological background

The Canary Islands are an archipelago of seven large and a few small islands in the Atlantic ocean, just off shore from the continental margin of Africa (Fig. 1.1). The nearest point on the African coast, Cape Juby, is just over 100 km east of Fuerteventura, but geophysical surveys and occurrences of fragments of oceanic **peridotites** and MORB-type **gabbros** as **xenoliths** in the lavas indicate that all the islands have been built up on oceanic crust. The ocean floor around and between the islands is 3000–4000 m below sea level; thus, the true heights of the volcanoes that form the islands are all in excess of 4–5 km. The largest and highest island, Tenerife, represents the tip of a cluster of volcanoes well over 7 km high; the present summit volcano, Teide, is in fact the third highest volcano on Earth, after Mauna Loa and Mauna Kea on Hawaii.

Like the Hawaiian Islands, the Canaries are a group of intraplate oceanic-island volcanoes dominated by basaltic magmatism. This very broad similarity suggests that they are also related to an upwelling **mantle plume** or **hotspot**, and there are many features of the geochemistry of the magmas erupted in the archipelago that support this in general terms. Independent geophysical evidence also points to the presence of a mantle plume in the area, albeit a much less vigorous one than the Hawaiian plume, most probably located close to the island of El Hierro. Although there are significant differences between the two groups of islands, these can be explained within the framework of a hotspot-type model when differences in the tectonic setting are taken into account.

Ages and spatial trends in age

The ages of the oldest exposed subaerial igneous rocks on each island are indicated in Figure 1.2 and in Table 1.1. In very broad terms there is an age progression from east to west, with Fuerteventura having the oldest rocks of the present hotspot activity (*c.* 22 million years ago according to the most recent **K/Ar radiometric age** determinations) and El Hierro the

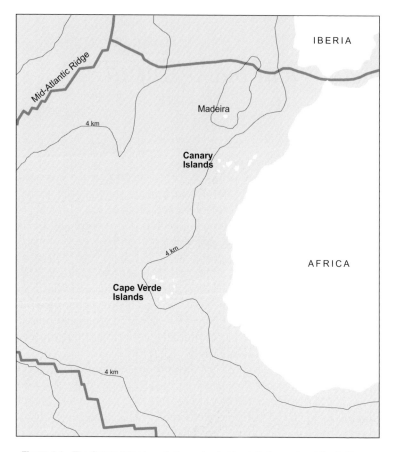

Figure 1.1 The Canary Islands and other volcanic islands in the eastern Atlantic Ocean.

youngest (1.2 million years ago). However, more recent activity is distributed throughout the archipelago, with only La Gomera lacking **Holocene** eruptions; the largest eruption in historical times actually occurred on the eastern island of Lanzarote and lasted from 1730 to 1736.

In broad terms, the pattern of activity in the islands follows that of

Table 1.1 Ages of main periods of volcanic activity in the Canary Islands. Ages in millions of years before present, based on radiometric dating.

Stage	Fuerteventura	Lanzarote	Gran Canaria	Tenerife	La Gomera	La Palma	El Hierro
Shield-building	22–11.8	15.5–5.0	14.5–8.0	12.0–4.0	11.0–4.0	2.0–today	1.2–today
Post-erosional	5.1–today	3.7–today	5.5–today	3–today	inactive	*	*

* Each of these islands is currently still in its main shield-building phase of activity.

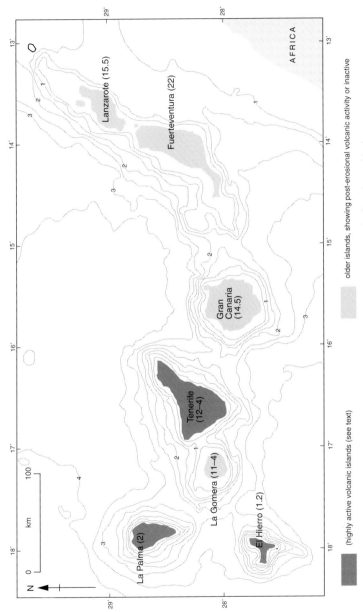

Figure 1.2   Ages and stages of growth of the seven main islands in the Canarian archipelago.

(highly active volcanic islands (see text))

older islands, showing post-erosional volcanic activity or inactive

Lanzarote (15.5)

Fuerteventura (22)

Gran Canaria (14.5)

Tenerife (12–4)

La Gomera (11–4)

El Hierro (1.2)

La Palma (2)

AFRICA

N

km

0   100

Hawaii. An early seamount stage, represented by rocks erupted below sea level and associated **intrusions**, composed of both coarse- and fine-grain intrusive rocks, forms the cores of the islands. Although erupted below sea level, these rocks have been exposed by uplift and erosion on Fuerteventura, La Gomera, La Palma and possibly on Tenerife. This is likely to be the most rapid and voluminous stage of activity in most oceanic volcanic islands, but it is only on these islands and in certain of the Cape Verde Islands (Fig. 1.1) that rocks produced in this stage of activity are exposed above sea level (see Chs 3, 7, 8). Once the islands have emerged, intense eruptive activity continues in the shield-building stage, named after the characteristic morphology of the large subaerial volcanoes produced at this stage. These shield volcanoes are characterized by eruptions from many flank vents organized into radial rift zones (normally three); but, unlike most comparable island volcanoes elsewhere, those in the Canaries commonly have steep slopes of up to 20° or more. Most of the islands contain two or more overlapping shield volcanoes; Tenerife has at least three, as a result of repeated shifts in the centre of activity on the island.

At the end of the shield-building stage, large summit-**collapse calderas** developed in some of the volcanoes, of Gran Canaria and Tenerife in particular. In the islands of Lanzarote, Fuerteventura, Gran Canaria and La Gomera, which first emerged above sea level between 22 and 10 million years ago, the shield-building stage has been followed by a period of volcanic quiescence and intense erosion of up to several million years, comparable to, but longer than, similar periods of inactivity that have been identified in the older islands of the Hawaiian chain. However, this quiescent period has ended on all of these islands; the other, older islands have experienced further volcanic activity. On Gran Canaria and Lanzarote in particular, this renewed activity has been much more intense than the "post-erosional" activity in Hawaii and has produced substantial volcanic sequences, over 1 km thick in places. The youngest islands, La Palma and El Hierro, which first emerged between 2 and 1 million years ago, are still very much in their shield-building stage of growth. Tenerife, whose oldest subaerial rocks are as much as 11–12 million years old, is more complex. It has long been thought to still be in its shield stage of growth, but recent work indicates that shield-stage activity ceased about 4 million years ago and that the more recent volcanic activity in the centre of the island represents a particularly vigorous post-erosional volcanism comparable to that of the Roque Nublo volcano on Gran Canaria. The long periods of continued activity after initial growth of each island are however matched by relatively low magma-eruption rates, so in volumetric terms the products of the early shield-building stage dominate the volcanoes as a whole, especially those parts below sea level.

In contrast to the pattern seen in the Canaries, the Hawaiian Islands show a very marked and simple age progression with distance from the mantle plume under Hawaii itself. In part, this difference is likely to reflect the much greater velocity of the Pacific plate over the Hawaiian plume: the northwestern part of the African plate is moving very slowly relative to the well defined underlying hotspots (such as the Cape Verde and Ahoggar hotspots). This may explain both the slower progression from seamount and shield to post-erosional stages of growth in the Canaries and also the larger volumes of the post-erosional volcanic sequences, as these islands have remained in the immediate vicinity of the hotspot for much longer.

Composition of magmas

In broad terms the early part of the growth of each island (the shield-building stage) is dominated by eruption of basaltic **ocean-island tholeiite** magmas (mildly silica-saturated to transitional basalts with moderate incompatible element contents), whereas later activity is usually more alkaline in character. The remarkable 1730–36 eruption of Lanzarote, which involved magmas of all compositions from **basanite** through **alkali basalt** to strongly silica-oversaturated tholeiitic basalt, is a unique exception to this rule. This broad pattern is similar to that encountered in Hawaii, but there is the important difference that on all the islands, but especially on Gran Canaria and Tenerife, the volcanism includes substantial **felsic** volcanoes erupting **trachyte** and **phonolite** magmas. As a result, large-scale explosive eruptions and collapse calderas have occurred on Gran Canaria and Tenerife.

The difference in the style of eruptions and the greater spread of magma compositions in the Canaries, as compared to Hawaii, most probably reflects differences in rates of supply of magma from the melting regions in the underlying mantle. Magma-supply rates during the shield-building stage in the Hawaiian Islands are so great as to prevent shallow magma chambers undergoing progressive fractional crystallization to produce felsic magmas, whereas in the later stages the supply rates are so low that any small shallow magma chambers that do form are starved of magma and solidify. In contrast, it seems that, for much of their history, Canarian volcanoes are characterized by moderate magma-supply rates that are high enough to keep shallow magma chambers molten while allowing fractional crystallization to occur.

In contrast to the large-volume eruptions of basaltic lava from the main Hawaiian volcanoes, most eruptions in the Canaries have been of relatively small volume, other than a few events in Lanzarote and Fuerteventura, and the rare large explosive eruptions of felsic magmas. The eruptions typically produce lava flows that do not travel far from the vents (kilometres, rather than tens of kilometres as in Hawaii), or phonolitic to trachytic **lava domes** in the case of most felsic eruptions. Furthermore, many of the basaltic eruptions are gas rich and produce large **scoria** and **spatter** cones around the vents. This concentration of eruption products near the vents has an interesting consequence for the morphology of the islands.

Morphology of the islands and subsidence histories

In contrast to the very gentle slopes characteristic of the Hawaiian shield volcanoes (typically less than 6° overall), the Canary Islands, with the exceptions of Lanzarote and Fuerteventura, have overall slope angles of 15–20° from their summits to well below sea level. This difference reflects the greater importance of viscous felsic lavas and scoriaceous mono-genetic volcanic cones, and also the smaller volumes of individual erup-tions in the Canaries as compared to Hawaii, as noted above. These greater slope angles, the semi-arid climate in many parts of the islands (which greatly reduces vegetation cover) and the lower eruption rates mean that erosion is an important factor in controlling the landscape, in contrast to the Hawaiian Islands where erosion is only significant in the post-shield stages of quiescence. Most of the islands, even on the active volcanoes, are characterized by abundant and often very deep gullies and canyons, known locally as **barrancos** and varying in depth from a few tens of metres to more than a kilometre. The barrancos separate plateau and ridge areas (the **cumbres**) on which erosion is much slower. Badlands-type terrain is often developed in the softer pyroclastic rocks.

Part of the reason for the gentler slopes on Lanzarote and Fuerteventura is that the older parts of these islands have been very deeply eroded, espe-cially in the Basal Complex of Fuerteventura, which includes uplifted deep-marine limestones, large alkali gabbro to **carbonatite** intrusions (representing solidified magma chambers) and a dense dyke swarm. Most of these rocks represent the seamount stage of the island's growth. The erosion in the Basal Complex is much deeper than in most oceanic islands, because Fuerteventura, unlike most others, underwent uplift for much of its early history, rather than subsidence (hence the outcrop of the deep-marine sediments). Similarly, Lanzarote shows evidence of subsidence only in the central **graben**, which contains its recent volcanic rocks.

Of the other islands, Gran Canaria appears to have undergone some subsidence after the **Miocene** phase of activity. More recently, in **Pliocene** and Quaternary times, it has been mainly stable, but with overall tilting perhaps accompanied by faulting: the north and east of the island have undergone slight uplift, whereas the south and west have subsided. It is only in the islands to the west, notably Tenerife, that evidence exists for significant subsidence of the earlier volcanic edifices, and even in these islands there is also evidence for uplift (notably on La Palma) associated with intrusion emplacement during the seamount stage. In all the islands, occurrences of early to mid-Quaternary beach deposits and other indicators of palaeo sea level close to present-day sea level suggest that in the past million years or so there has been relatively little uplift or subsidence.

The subsidence histories of the Canary Islands are therefore in marked contrast to those of the Hawaiian Islands (and, indeed, those of most oceanic islands). Several factors may be involved. First, the slow-moving hotspot means that thermal subsidence (as the lithosphere cools on moving away from a hotspot) is not significant, as it is in the older Hawaiian Islands. Secondly, the Canary Islands have been emplaced on some of the oldest, coldest and thickest **oceanic lithosphere** in the world (of Jurassic age, 165–176 million years ago). Such lithosphere, if undisturbed, is also very strong and well able to support the weight of the volcanoes. Although there are indications that the lithosphere beneath the western islands has been heated and weakened by the hotspot, that to the east may have retained much of its strength. Apart from intense erosion and moderate eruption rates, the other significant factors controlling the morphologies of the islands are volcano structure and the occurrence and geometry of lateral collapses, which produce very large collapse scars extending up to the summits of the islands, as discussed below.

Controls on volcano structure

Because most volcanic islands are emplaced on oceanic lithosphere, which forms rigid plates, they show structures dominated by the deformational effects of the volcanoes themselves. In essence, the volcanoes deform under their own weight and as a result of injection of pressurized magma into them. Thus, radial **dyke** swarms and concentric **inclined sheets** (**cone sheets**) are common, as are triple or "Mercedes star" rift-zone structures: apart from instances in the Canaries, these are exemplified by Mauna Kea on Hawaii and Piton des Neiges on Réunion. Departures from this structure, such as Piton de la Fournaise (Réunion) and Kilauea (Hawaii) are most usually attributable to volcano growth on the flanks of an older volcano, which acts as a supporting buttress: as the growing volcano starts to slide down the flank of this older volcano, one of the arms of the triple rift system is suppressed in favour of the others.

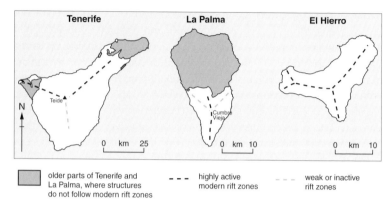

Figure 1.3 Three-arm volcanic rift zones in the modern volcanoes of Tenerife, La Palma and El Hierro.

The highly active islands of Tenerife, La Palma and El Hierro show Mercedes star distributions of Quaternary volcanic vents, with the different arms of the rift zones developed to varying degrees (Fig. 1.3) especially on La Palma and Tenerife: evidence for possible causes of the uneven development of the rifts on these two islands can be seen in the field. In the case of Tenerife in particular, there has been much controversy over the significance of recent volcanic vent distributions, with the alternative suggestion that the growth of the island has been controlled throughout its history by dyke emplacement along northwest and northeast structural lineaments inherited from the pre-existing oceanic lithosphere and reactivated by regional deformation. Similar triple-rift Mercedes star geometries have also been identified in the older islands, especially Fuerteventura, on the basis of dyke-swarm trends, but these are less evident in the topography because of the effects of erosion and the complete removal of parts of these older edifices by erosion and lateral collapse. Furthermore, some other trends have been identified. NNE–SSW structural trends dominate in the older parts of Fuerteventura and Lanzarote, although Quaternary activity in Lanzarote is concentrated along ENE–WSW-trending fissures in an E–W trending graben (or pull-apart structure). The dominant structural trend in Gran Canaria seems to be orientated NW–SE, although the later Miocene volcanism is dominated by a subcircular caldera structure (see Ch. 5) and a radial dyke swarm is present in the Pliocene Roque Nublo **stratovolcano.**

Volcano lateral collapse structures

Volcanoes are composed of relatively weak unconsolidated rocks (especially if a significant proportion of pyroclastic rocks are present) and are subjected to a variety of stresses, from intrusions into the volcanic edifices, from earthquakes caused by the weight of the volcano and by regional deformation, from the surrounding sea water (in the case of island volcanoes) and from movement and thermal expansion of groundwater and hydrothermal pore fluids within the volcanic edifice. As a result, they often collapse in catastrophic events. There are two principal types of collapse. First, central or vertical collapses caused by eruption or lateral drainage of large volumes of magma from shallow magma chambers produce central collapse calderas; the largest examples are associated with felsic magma chambers and include the Miocene Tejeda caldera of Gran Canaria and the younger calderas of the Cañadas volcano on Tenerife. It should be noted that "caldera" has a morphological as well as structural meaning (see Glossary) and that in Spanish-speaking countries any large depression on a volcano, especially near its summit, will be called a caldera, although it may not have been produced as a result of magma-chamber evacuation; examples of the confusion that this may cause include the Caldera de Taburiente on La Palma and the Caldera de las Cañadas on Tenerife.

In recent years, especially after the May 1980 eruption of Mount St Helens, it has been recognized that **lateral collapse** or landsliding of the flanks of volcanoes is a common, volcanologically very important and potentially extremely hazardous process. The largest terrestrial landslides, with volumes of up to 5000 km³, have been identified around oceanic-island volcanoes, and those in the Canaries are second only to those around Hawaii in number and volume (Fig. 1.4). Unlike the Hawaiian collapse structures, which are mostly submarine, those in the Canary Islands mostly extend well up into the subaerial parts of the pre-existing volcanoes, producing spectacular collapse scars up to 20 km across and 2 km deep.

The Canaries offer the opportunity to examine lateral collapse structures in various stages of development, from old and deeply eroded (the Roque Nublo structures on Gran Canaria), through relatively recent and still morphologically well defined, although often partly filled in and with significant erosion of the headwall (Taburiente and Cumbre Nueva, La Palma; Icod, La Orotava and Güímar, Tenerife), to incipient lateral collapse structures, in particular the Cumbre Vieja on La Palma.

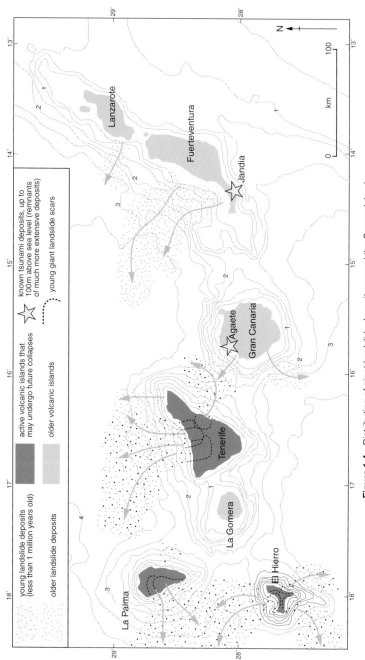

Figure 1.4 Distribution of giant landslide deposits around the Canary Islands.

Legend:

young landslide deposits (less than 1 million years old)

older landslide deposits

active volcanic islands that may undergo future collapses

older volcanic islands

known tsunami deposits, up to 100m above sea level (remnants of much more extensive deposits)

young giant landslide scars

Lanzarote

Fuerteventura

Jandia

Gran Canaria

Agaete

Tenerife

La Gomera

El Hierro

La Palma

N

km

0   100

Eruptions recorded in documents

Frequent but mainly small-volume and basaltic eruptions have occurred on three of the islands since the Spanish conquest:

Tenerife:

1704, 1705, 1706	Phases of one extended eruption over several parts of the island; one phase destroyed the town of Garachico.
1798	Last eruption of Pico Viejo.
1909	Chinyero eruption.

An extended period of earthquake activity, beginning in 1989 and lasting until 1992, was probably related in part to shallow **intrusive** and **extrusive** activity on the floor of the ocean between Tenerife and Gran Canaria. Early accounts of eruptions in 1430 and 1492 are now regarded as erroneous.

Lanzarote:

1730–36	The second-longest and third-largest basaltic eruption in historical times.
1824	1824 Tao, Nuevo andTinguatón eruptions.

La Palma:

1585	A very complex eruption at Jedey, which also involved the emplacement and near collapse of a phonolite **cryptodome**.
1646	Eruption on the flank of the prehistoric Volcán Martín.
1677	Fuencaliente eruption.
1712	El Charco eruption.
1936	A possible submarine eruption that caused extensive earthquake damage on the southern tip of the island.
1949	Eruption from several vents with faulting and movement of the western flank of the Cumbre Vieja volcano (Ch. 8).
1971	Teneguía eruption.

An earlier eruption is thought to have occurred in 1480, just before the arrival of the Spanish; this has been confirmed by radiometric dating.

Prehistoric but geologically very recent volcanic activity is more varied and includes the highly explosive (sub-**Plinian**) pumice eruption of Montaña Blanca on Tenerife in about 70 BC, and dome collapses and block-and-ash pyroclastic flows on La Palma. These eruption types are considerably more hazardous than the basaltic lava flows, Strombolian scoria eruptions and minor hydrovolcanic explosions which have occurred in historical times.

Chapter 2

Environment and logistics

Climate, flora and fauna

The Canaries lie close to the Atlantic margin of the Sahara, but except on the comparatively rare occasions when southeast winds bring hot, dry and dusty air ("tiempo sur") across the islands, this is not apparent except in the eastern islands, especially Lanzarote. The climate of the other islands is generally very pleasant and mainly non-seasonal. However, there are substantial variations according to height and wind direction. These result in a weather pattern typical of oceanic islands but very different from that on the continents; it is worth bearing this in mind, as it can affect plans for excursions and it also has a profound influence on the flora of the islands.

Most of the time, moderate northeasterly trade winds bring cloud and sometimes light drizzle to the north and east of the higher western islands, as warm moist air is blown up the windward slopes (Fig. 2.1). As the air rises, it cools and the moisture condenses out, forming **orographic clouds**; once the air passes over or around the summits, condensation ceases, so this weather pattern leaves the south and west of these islands sunny and

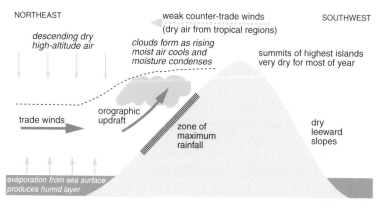

Figure 2.1 Wind and weather patterns around high oceanic islands in the trade wind zones, such as the Canaries.

dry. The cloud is generally least extensive early in the morning and late in the afternoon and evening, clearing altogether at night. It can thicken rapidly and annoyingly during the morning and in the middle of the day, so in the higher islands it is often well worth getting out into the field at the crack of dawn. Under some conditions an extensive layer of cloud, the Mar de Nubes, forms out over the ocean, with the peaks of the islands projecting up through it. Less frequent southwesterly and westerly winds, often of gale force, reverse this pattern. Every so often in autumn and winter, westerly to northerly winds cover the whole of the islands with cloud and rain. If the wind is northerly, freezing rain and snow can fall on higher ground in winter, and blizzards are not unknown above about 1500 m.

When clouds are present, they mostly occur between 800 m and 1800 m, with warmer and much drier air above what is known as the trade-wind inversion. The temperature change across the inversion can be startling, with upward temperature increases of as much as 5°C over a vertical range of a few hundred metres. The air above the inversion is in fact trade-wind air that has flowed as far as the tropics, where it rises, loses its moisture in thunderstorms and hurricanes, and flows northwards again in a reverse flow that is sometimes known as the counter-trades, although these winds are much weaker than the trade winds.

The result is a strong vertical zonation in the vegetation. Below 800 m, in the south of each island, semi-arid to desert conditions prevail: scrub vegetation, prickly-pear cactus and spurge dominate. More European types of vegetation occur in the north and east of each island at this level, and this is also where most cultivation occurs (winter salad vegetables, pineapples and bananas) and where most people live, so little natural vegetation remains. Between about 800 m and 1400 m, in uncultivated areas, laurel and tree heather (up to 5–8 m high) form thick and impenetrable woods, especially on the wetter sides of the islands; almost equally impenetrable Mediterranean maquis and scrub occurs at this level on the drier sides of the islands. The most pleasant vegetation occurs between about 1400 m and 1800 m: the cloud forests of Canary pines. These trees show remarkable adaptation. Being close to the cloud tops, they receive relatively little rain but have needles up to 15 cm long, on which water droplets from the clouds collect and drip onto the ground. The Canary pines are the gardener's dream: self-watering plants. Higher still, above the cloud tops for most of the time, the pines are replaced by scattered junipers and dry alpine scrub; although these areas are among the driest parts of the islands, succulent plants are restricted by the intense cold at night (frosts are very common here). Above 2000 m, bare rock is widespread.

The Canary Islands have a variety of native plants, notably the tree heathers and the pines. Others include the celebrated dragon trees: giant liliaceas that occur only outside the archipelago on São Nicolau in the Cape Verde Islands and on the island of Socotra, off Somalia, in the Indian Ocean. It seems that these plants once occurred all across North Africa, but became extinct elsewhere with the development of modern grasslands and the desiccation of the Sahara. Native animal species are less obvious and have suffered more from the introduction of competitor species by human beings. The giant lizards of Gran Canaria and El Hierro are now very rare, although smaller lizards and skinks remain common in drier areas. Bird species include near-flightless pigeons (also now very rare) and the more successful Canarian hawk or Barbary falcon, a distinctive rust-coloured bird that occupies a niche similar to that of the kestrel. More familiar bird species present in the islands include buzzards, ravens and the spectacularly fast swifts. The most noticeable insects are dragonflies and millipedes, some of which are very large, and locusts blown across from Africa during periods of southeast winds. Ticks, midges, snakes and other poisonous animals do not occur; the only dangerous animals in the islands are bees and guard dogs.

Getting to and around the islands

Although the islanders themselves may be ambivalent about the tourists, there is no doubt that the tourist trade makes it much easier to get to the Canaries from all parts of Europe and it also facilitates other aspects of geologically orientated visits to the islands. Cheap tickets on charter flights are widely available to Tenerife and Gran Canaria in particular, the only notable restriction being that most tickets restrict stays on the islands to two weeks. For longer stays, or for visitors from outside Europe, the best plan is to travel to Madrid or Barcelona and take an internal flight from these cities to the islands; flights from the Spanish mainland serve all the islands except La Gomera and El Hierro, although again the most frequent flights are to Tenerife and Gran Canaria.

Travel between islands can be by air, principally on the regional airline Binter Canarias (originally a subsidiary of the Spanish national airline Iberia, so tickets can be obtained through Iberia ticket offices if travel agents cannot oblige), or by sea, on the ferries of Trasmediterranea or Fred Olsen Ltd. Apart from hydrofoil and fast ferry services between Tenerife and Gran Canaria, ferry voyages take several hours or are overnight. Both ferry services and flights are organized around the hubs of Tenerife and Gran Canaria; direct flights from the smaller eastern islands to the smaller western islands are rare, and ferry services non-existent. In general, air travel is more convenient, although more expensive. However, it should be borne in mind that Tenerife has two airports, Reina Sofia in the south and Los Rodeos in the northeast, and although most international flights arrive at Reina Sofia the inter-island flights all operate from Los Rodeos. Allow one hour (by taxi) or two to three hours (by bus) to travel between the two, and aim to arrive in good time for any Binter Canarias flight to avoid problems with overbooking. A further problem to bear in mind is that flights to and from La Palma and El Hierro (and occasionally also the ferry services) can be disrupted by bad weather, especially in winter. It is often a good idea to aim to spend a couple of days in Tenerife or Gran Canaria at the end of visits to these islands, to provide some leeway in travel arrangements.

A few light aircraft and helicopters are available for hire (at a high price), but aerial tourism in the Canaries is not developed to anything like the same extent as it is in Hawaii or Alaska. This is partly because the weather is something of a lottery: if you do decide to splash out (and, if you are lucky, the views – especially on Tenerife and La Palma – can be well worth it), try to organize an early-morning flight. Another reason for arriving early for Binter Canarias flights is that you stand a better chance of picking the best seat for the views from the planes, although there are no seat reservations. In most cases, the best views are from the port side flying west and starboard side flying east.

On the islands themselves, the abundance of tourists also ensures an abundance of hire cars. These are generally very cheap (but remember to check that insurance is included when comparing prices) and are the best way of getting around unless you are a committed walker and prefer to use the bus services. These are of widely varying frequency, but are cheap, unlike taxi services outside the main towns. Hire cars cannot be taken between islands, and most companies also exclude off-road driving (even for four-wheel-drive vehicles). Do not forget your driving licence (and those of any alternate drivers in your party) and a credit card; all hiring drivers must be over 21 years old. Four-wheel-drive vehicles are available for hire (at a high price), but, with the possible exception of southern Gran Canaria and the forest tracks of the western islands, the roads and tracks are either good enough to be passable in ordinary cars, or are non-existent. Severe restrictions on off-road driving apply in the national parks on Tenerife, Lanzarote and La Gomera, and in other protected areas (see below). All the itineraries described in this book involve on-road driving only, but the roads in the interiors of the islands are often steep and winding. If you have a choice of cars, take the smallest one that will comfortably accommodate your party, as it will mean less work for the driver.

Practicalities in the field

Logistics

Details of shop-opening times, mealtimes and other practicalities can be found in general guides to the islands (p. 284). One point to bear in mind is that shops open at 08.00 h, although some hotels and many cafes serve breakfast before then; so if you want to be out early to see the dawn or catch the best of the weather, stock up the night before on food and drink for lunch. Many shops are closed on Sundays and some hotels and restaurants close on Mondays, so plan ahead before those days in particular. Later in the day, shops often close between 14.00 h and 16.00 h, but then open again until 20.00 h, so for much of the year your time in the field is limited by the light rather than by the need to stock up. Similarly, most restaurants stay open very late, especially at weekends.

Bottled water is plentiful and cheap. The tap water is generally safe, but is best at high elevations and on the windward slopes; in the small villages it is often piped directly from the galerías (water tunnels) that have been excavated to extract groundwater. On the coast, and in the leeward slope resorts in particular, the water is heavily treated, sometimes produced by desalination, and is safe but has an unusual taste. Problems from restaurant food are rare and are usually associated with salads and unpeeled fruit.

Negative colour film is widely available, transparency film less so, and specialist films can be found only in the major towns. Take plenty of film with you, as there are many opportunities to take panoramic mosaics from the vantage points noted in this guide. The vagaries of the weather mean that photography in the islands can be frustrating, but if you get it right the results can be superb. The best light, especially for landscapes, is generally early in the morning and in the evening; during the day, haze and the sheer intensity of the sunlight can combine to wash out most pictures on the one hand or create excessively intense shadows on the other.

National parks and other restricted areas

About 42 per cent of the area of the Canaries is protected in one way or another. At the top of the scale are the four national parks run by the Ministerio de Medio Ambiente (Parques Nacionales): Timanfaya on Lanzarote, Las Cañadas del Teide on Tenerife, Garajonay on La Gomera and Caldera de Taburiente on La Palma. Access in all of these is restricted, with visitors confined to marked paths and roads (itineraries in the parks described here follow these entirely). Absolutely no collecting of rocks or plants is allowed without an official permit. Next down on the scale are regional parques naturales, for example the areas around Pajonales and Roque

Nublo on Gran Canaria, the crest of the Dorsal de La Esperanza on Tenerife, and the crest of the Cumbre Vieja on La Palma. These are also policed (especially along the roads) and the rule is that no collecting is allowed unless you have a scientific investigations permit. On a more local scale, there are various wildlife and plant preserves (of which perhaps the most famous is the UNESCO world biosphere reserve at Bosque de Los Tilos on La Palma), sites of scientific interest, and archaeological sites. These are generally signposted and again are hammer-free zones – especially in archaeological sites, in respect of the fact that the most characteristic relics left by the Guanches were rock carvings or petroglyphs.

Safety
A good starting point for discussion of safety are the sets of rules for walkers in the admirable walking and hiking guides to the islands, written by Noel Rochford (see p. 284) and the Geologists' Association guidelines for fieldwork. Some points specific to the Canaries or that require particular emphasis are:

- Take care to avoid sunstroke and sunburn, especially at high altitude. Carry plenty of water (a minimum of a litre per person for a full day in the field, and two or three litres in hot weather, when it is also advisable to have isotonic drinks or rehydration packs in reserve). Put on plenty of high-strength sun lotion and keep putting it on at intervals during the day (with sunblock in reserve); wear shirts and long trousers rather than shorts; and wear the largest hat that you can find, with a chinstrap or string to keep it on in the wind.
- Even if the day begins sunny, do have windproof, waterproof and warm clothing in your pack, especially if the wind is from the west or north. If it rains, stay out of the barrancos, as they are prone to flash floods.
- Wear proper hiking or walking boots, because the going underfoot is often very rough. Take special care when walking over rough lava, scoria or blocky rubble, as it is prone to shift under foot. Stick to the paths as much as possible in this sort of terrain, and make sure your first-aid kit is well packed with materials for patching up cuts and scrapes.
- Do not light fires, except in the stone or concrete barbecue stands at certain campsites, and do not smoke. This rule is imperative in the pine forests, where major fires commonly occur. If there is a fire, stay well away (especially downwind), as it can spread at life-threatening speed. Certain pine forest areas are closed altogether in very dry weather.
- The main shooting season (for rabbits and birds) is in October and November; when hunters are about, be careful to remain visible at all times, especially if you are off the main paths.
- Be careful to follow known paths and routes, and leave with hotel

owners or other responsible parties details of where you are going. Make sure you have maps and a compass with you, and do not hesitate to backtrack if you get lost, especially if the cloud comes down. Avoid scrambling down steep slopes or crags; volcanic rocks are especially treacherous. Be careful as well on slopes covered with pine needles, as these are prone to slide away underneath you.

- Stay away from the inner rims of recent craters (do not even think about climbing down inside them except on well established paths such as that into the Caldera de Bandama on Gran Canaria) and other unstable cliffs.
- In areas where rock sampling is permitted (see above for restrictions), be especially careful when hammering: many of the rocks are very brittle.
- Do not go into lava tubes or open water tunnels (galerías) unless you are and experienced caver, have left detailed information on where you are going, and are accompanied by a member of one of the various speleological clubs in the islands. Gas seepages into the tubes and tunnels can cause asphyxiation.

Having said all of the above, do not let these points put you off. So long as you follow these and the other normal guidelines, you are far more likely to encounter problems on the roads than off them. In this regard, be especially careful when driving in the cities and on the winding mountain roads in Gran Canaria, Tenerife and La Palma.

Car crime

With one exception, crime levels in the Canaries are generally low, except in the more dubious areas of the major cities. The exception is that of break-ins to hire cars, especially those left in small parking places at the side of the road by walkers. Even if nothing is stolen, there is nothing more annoying than losing part or all of a day returning a damaged car to the hire company. The only way to avoid this is not only to leave nothing in your hire car (except maybe your driving shoes and a spare bottle of water), but **to make it obvious that there is nothing there**. So leave the glove compartment open and, if you have a hatchback, take out the back shelf and leave it on the seat as well.

Accommodation[1]

The easiest and often the cheapest accommodation option, especially on the main holiday islands of Lanzarote, Fuerteventura, Gran Canaria and

---

1 See p. 284 for websites containing details of places to stay and contact addresses for the island tourist offices through which bookings can be made for casas rurales and other local options.

Tenerife, is to book a hotel room or apartment as part of a package-holiday deal. The downside of this is that you will more often than not end up staying in some soul-less resort from which it will be a pleasure to escape each morning – preferably early, before the coach tours start to clog up the roads – and where it is difficult to find good food in the evenings. More pleasant alternatives are:

- Hotels and apartments in the major towns and cities. These are often very pleasant, with much character, and there will be plenty of choice for restaurants in the evening. The downside is that you will be returning through heavy traffic at the end of the day, and finding parking space will be a chore. If you are a light sleeper, you are likely to find the street noise irritating at night, especially at weekends, as it goes on into the small hours.
- Hotels in the interiors of the islands are more plentiful as you go west. They fall into two main categories: the paradores nacionales, which are luxurious but remote and wildly expensive, and small hotels and pensiones in the hill towns and villages. The latter are often what the French call restaurants avec chambres: they cater to the tourist trade during the day and have a few rooms upstairs. These are often basic but are cheap and can have superb locations.
- The most interesting self-catering option is to stay in one of the casas rurales, converted cottages or farmhouses that are the rough equivalent of the French gites and are organized on an island-by-island basis with bookings through the tourist offices on each island. They score highly on character, are excellent value, but are not luxurious.
- If you want to camp, there are official campgrounds, but camping elsewhere is frowned upon and is illegal in many places, especially in protected areas. Some of the official campgrounds – for example, that in the centre of Caldera de Taburiente – impose strict limits on the duration of your stay. There are few places that are not accessible in a day trip from the towns (the deep interior of Caldera de Taburiente is the principal exception). Stay in a pension or casa rural is preferable.

You do not need to speak Spanish, but it helps
Most people with either a commercial or official interest in the tourist industry, ranging from hire-car company clerks through the staff of large hotels to many policemen, speak at least a little English, although older people are more likely to have learned French in school. An important exception to this rule commonly occurs at check-in desks for inter-island air and ferry services, although, if you are really stuck, the supervisor or reservations clerk is likely to be able to help. Bus and taxi drivers vary widely in their linguistic abilities, and an ability to speak Spanish is

definitely useful if you are staying in a small town or rural area. Further-more, you are likely to have better service and a friendlier response from the locals if you have at least the rudiments of Spanish. A few language classes or sessions with a language book or CD-ROM before you go are well worth the effort, and the more you know the more you will get out of your visit. On the other hand, in the resort ghettoes one wonders if it would be more useful to speak Scouse, Geordie, German, Dutch or Swedish, according to the dominant source of visitors.

Chapter 3

Fuerteventura

Introduction

Fuerteventura is the second largest of the Canary Islands (1662 km^2) and also the longest, at over 100 km (Fig. 3.1). It is also only 100 km from the continent of Africa, the closest of the islands. Like its close neighbour to the north, Lanzarote, it is located upon oceanic crust. Fuerteventura rises some 3000 m from the ocean floor to the sea surface, and 807 m more to its highest point, Pico de La Zarza on the Jandía Peninsula in the south of the island. However, most of the island is less than 200 m above sea level, which reflects intense erosion since the principal shield-building phase of volcanism. Fuerteventura has extensive exposures of the cores of the volcanoes that built up the island, including the uplifted remains of the seamount upon which it subsequently grew. As noted in Chapter 1, the Canaries are among the best places in the world to examine the evidence for this stage in the growth of oceanic-island volcanoes. In Fuerteventura it is even possible to see uplifted slivers of the ocean-floor sediments upon which the seamounts grew, as well as a variety of much younger shallow-water marine sediments that have accumulated upon the flat eroded surface of the island during periods of high sea level.

The principal elements of the geology of Fuerteventura are indicated in Figure 3.1. Besides the marked elongation of the island, it is notable that the oldest rocks are mostly exposed along the western coast, with younger rocks to the east. This suggests that nearly half of the original volcanic edifice has been removed, including the summits of the main shield-stage volcanoes. Although more recent subaerial erosion has almost completely removed evidence for landslide scars, except at Jandía in the south, other evidence exists for the occurrence of giant lateral collapses during the shield stage of activity on Fuerteventura, as on the younger islands. Marine geological surveys indicate that giant **landslide deposits** are located off the western side of the submarine volcanic ridge on which the island is located (see Fig. 1.4). A broad shallow (less than 200 m deep) submarine platform is present around the island, implying that the size of

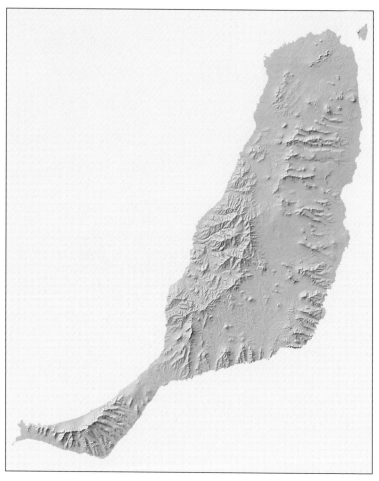

Figure 3.1 Shaded-relief topographic image of Fuerteventura. Note in particular the Betancuria Massif in the centre-west of the island, surrounded by the Central Depression, and the northwest-facing scarp of the Jandía peninsula in the south. Courtesy GRAFCAN.

the island has been further reduced by more recent coastal erosion. This platform encloses both Fuerteventura and Lanzarote, and the strait between the islands is as little as 40 m deep; thus, from a geological point of view they form one large elongated volcanic edifice. During periods of glaciation and low sea level in the Quaternary they formed a single island.

The rocks themselves can be divided into five main units, not all volcanic, as indicated in Table 3.1. These correspond to the units indicated on Figure 3.2, except that the Jurassic to Cretaceous deep-marine sequence is included in the Basal Complex with the seamount-series rocks on the

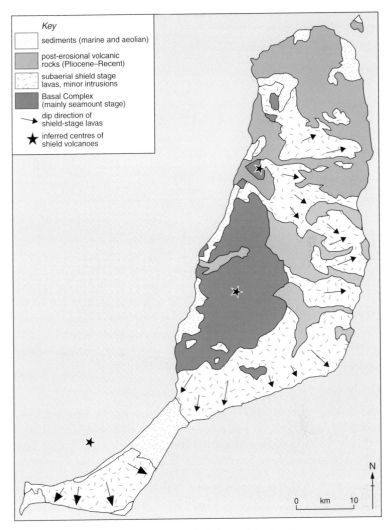

Figure 3.2 Simplified geological map of Fuerteventura.

map, as these earlier rocks occur only as small fragments or **screens** between intrusions.

The term "Basal Complex" was originally applied in the belief that the pre-shield submarine volcanic and intrusive rocks on Fuerteventura represented an uplifted fragment of the ancient crust upon which the island volcanoes grew. A later variant of this interpretation emphasized the intense dyke swarm and compared the sequence to uplifted fragments of oceanic crust or **ophiolites**, found in the Alps, Cyprus and elsewhere and

23

Table 3.1 Principal stratigraphic units in Fuerteventura.

Unit	Age	Rock types
Post-erosional volcanic rocks	5 million years to Recent	Alkali basalt lava flows and scoria cones
Sediments	Pliocene to Recent	Pliocene marine limestones; alluvium derived from volcanic rocks and aeolian sands derived from offshore
Shield-stage volcanic rocks	22–21 million years to 12 million years	Mainly basaltic lava flows, scoria and dykes; minor felsic rocks; youngest major intrusions in Betancuria Massif*
Seamount series*	48 million years to 22–21 million years	Submarine pillow basalts and hyaloclastites; reworked sediments including reef limestone breccias; dyke swarm and many major intrusions
Deep-ocean sediments and oceanic crust*	Early Jurassic (180 million years) to early Cretaceous	Pelagic limestones, and mudstone to siltstone detrital sediments, overlying mid-ocean ridge basalts

* Rocks included in the Basal Complex: see text for discussion.

interpreted as the products of seafloor spreading at mid-ocean ridges. However, subsequent investigations have shown that most of the rocks in this unit are younger than the deep-marine sediments and fragments of the ocean floor, and are much closer in age to the subaerial volcanic rocks, hence their reinterpretation as a seamount formed before the growing volcano emerged above sea level. However, "Basal Complex" remains useful as a local descriptive term on Fuerteventura because of the presence of the Jurassic–Cretaceous sequence and also because many of the younger intrusions are feeders to the subaerial shield volcanoes.

Jurassic–Cretaceous sequence
As noted above, the Jurassic–Cretaceous rocks are restricted to a small area within the Basal Complex and are intensely intruded by dykes and other intrusions of the seamount. However, the best exposures, close to the coast west of Betancuria, have been carefully mapped and a sequence of deep-ocean sediments defined (stops 3.9–3.13). These vary from limestones and chalks to sandstones, deposited as part of a deepwater submarine fan that grew out from the margin of Africa. The rocks at the bottom of the sequence are early Jurassic basalts; unlike every other volcanic rock in the Canaries, these have the geochemical features characteristic of mid-ocean ridge basalts. They appear to have formed at the Mid-Atlantic Ridge during the very earliest stages of the opening of the Atlantic Ocean.

The whole sequence becomes younger from south to north as it is tilted to near vertical and actually overturned in places. This intense deformation pre-dates all the younger rocks and is believed to be mid-Cretaceous in age. Offshore seismic and other surveys associated with the search for oil along the coast of West Africa have established the occurrence of an episode of intense deformation in the mid-Cretaceous period all along the continental margin from Morocco to Gabon, apparently related to uplift of the African continent. This led to tilting of the margin sediments towards the ocean and slow-moving but colossal slumps, cutting right down to the base of the sedimentary sequence at a depth of several kilometres, which displaced the margin sediments towards the ocean. The structures in the Mesozoic sediments on Fuerteventura may represent part of the intensely deformed toe of one such slump, most probably towards the side of the slump, as the rocks are tilted towards the north.

Recognition of the Mesozoic sedimentary sequence on Fuerteventura, and the fact that it is intruded by the rocks of the seamount series (including the dyke swarms), was critical to establishing that the latter are *not* part of the original oceanic crust: ocean-floor sedimentary sequences accumulate on top of, and are always younger than, the oceanic crust on which they form.

Seamount series
The seamount series forms most of the outcrop of the Basal Complex, which covers about $300\,km^2$ in west-central Fuerteventura. This area of relatively high ground, the Betancuria Massif, owes its existence to the high resistance to erosion of many of these rocks. The principal elements of the seamount series are:

- *Basaltic pillow lavas erupted on the ocean floor at great depth, probably as much as 3 km initially* These are of alkali basalt composition, distinct from the older mid-ocean ridge basalts and similar to the other hotspot-related rocks of the Canaries.

- *Volcaniclastic rocks (including **hyaloclastites** and re-sedimented breccias composed of fragments of pillow basalts and intrusive rocks), with intercalated sediments including calcareous debris* These are mainly exposed around the margins of the seamount series and they represent only a very small fraction of the total sequence. Where field relationships can be observed, the pillow-lava sequences grade upwards into these rocks, suggesting that they form a shallowing-upward or shoaling sequence developed as the seamount grew towards the surface of the ocean. The calcareous debris includes shallow-water sediment and coral and algal fragments. These suggest that, as the volcanic island emerged, it was fringed by shoals and reefs.

- *Very abundant dykes and fewer of more gently dipping sheet intrusions, including sub-horizontal sills* These mainly trend north-northeast and in the central region of the Betancuria Massif form more than 80 per cent of the outcrop over wide areas; percentages of dykes decrease to east and west, and are also particularly low in the extreme south of the seamount-series exposure. Most of these sheet intrusions are basaltic in composition, but some are trachytic, suggesting that they originated from the magma chambers represented by the major intrusions. A few minor intrusions have unusual compositions, including carbonatite (carbonate-rich rather than silicate-rich igneous rocks).

• *Major intrusions* The oldest of these are cut by almost as many dykes as the surrounding pillow lavas, indicating that they pre-date the dyke swarms, but the youngest post-date almost all the dykes and are likely to be coeval with the subaerial shield volcanoes (see below) rather than being part of the seamount series proper. The intrusions range in composition but are mainly gabbros and **pyroxenites**. The latter represent the products of crystallization from particularly hot magmas that rose directly from the mantle, with little cooling and crystallizing en route.

The older rocks in the seamount series, along with the Jurassic and Cretaceous rocks, show evidence of metamorphism and alteration as a result of heating by later intrusions and by reaction with circulating heated sea water. Locally, around the margins of the larger pyroxenite intrusions, the rocks were heated enough to undergo partial melting, which has produced some remarkable textures and structures (see stop 3.14). The younger rocks show progressively less alteration, but even the youngest rocks were altered by reaction with circulating rainwater during the shield-building stage.

This alteration has made it difficult to obtain accurate radiometric ages for the seamount-series rocks. The ages obtained range from 48 million years, in the oldest intrusions to around 20 million years in the case of the intrusions coeval with the subaerial shield volcanoes. This suggests that the seamount grew only very slowly. However, the fossils in the marine sediments are Oligocene (around 35 million years old) to early Miocene (around 22 million years old) in age.

A final important feature of the seamount series is its relationship to the overlying subaerial lavas of the shield-building stage. The two sequences are everywhere separated by an unconformity, which in places cuts deep into the intrusive core of the seamount, suggesting the removal of between hundreds of metres and a few kilometres of rock during the period between growth of the seamount and the start of extensive subaerial volcanic activity. Furthermore, this deep erosion took place in a very short time interval, since there is no great time gap between the youngest submarine sediments and seamount intrusions on the one hand and the oldest subaerial lavas on the other.

These observations, made at places described in the itineraries, provide important insights into how the transition from seamount to volcanic island has taken place in the Canaries. When eruptions take place in shallow water or through water-saturated sediment, boiling of water heated

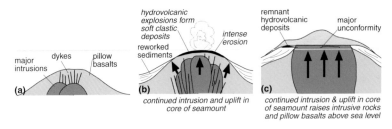

Figure 3.3 Emergence of a seamount above sea level in the face of intense erosion, by uplift resulting from the emplacement of intrusions within the seamount. **(a)** Seamount grows towards surface by growth of intrusions and eruption of pillow basalts. **(b)** As seamount grows to within a few hundred metres of the surface, intense erosion removes hydrovolcanic explosion deposits; core continues to grow by intrusion. **(c)** Island emerges as resistant core of seamount is raised to sea level, forming stable platform for eruption of subaerial shield-stage lava flows

by the magma causes the eruptions to become violently explosive and produce highly fragmented pyroclastic rocks. These hydrovolcanic rocks are commonly preserved when erupted onto more solid rocks such as subaerial lava flows, as tuff cones or **tuff rings** (see Ch. 4). However, when a volcano is first emerging above sea level, these solid substrates are not present and the first hydrovolcanic cones to emerge above sea level are rapidly eroded by the waves; historical examples include Graham Island in the Mediterranean (formed in 1831, but completely eroded by early 1832) and Metis shoal in Tonga, which has repeatedly emerged and been eroded again. The combination of intense wave action and soft hydrovolcanic rocks means that a volcanic island will generally not become fully established until it is covered by a carapace of lavas, as has happened in the case of Surtsey, off Iceland, formed in 1963. The geology of Fuerteventura suggests an alternative mechanism (Fig. 3.3).

The presence on Fuerteventura of fragments of the pre-volcanic ocean-floor sediments, and even of mid-ocean ridge lavas exposed above sea level, implies at least 3–4 km of uplift. As the islands lie within the interior of the African plate and the oceanic crust around them is undisturbed, the most likely uplift mechanism is emplacement of large volumes of intrusive rocks, resulting in swelling of the submarine volcano and uplift of the rocks above the intrusions (Fig. 3.3a). These intrusions were most probably those that fed the submarine volcanism and also the very large volumes of intrusions exposed at the present level of erosion in the Betancuria Massif. Thus, the continued intrusive activity, as the top of the submarine volcanic edifice approached the sea surface, would lift rocks originally erupted at greater depths (pillow basalts), as well as intrusive rocks, into the zone of intense erosion at sea level. Although some of these rocks would be eroded (hence the presence of their clasts in the re-sedimented breccias), their greater resistance to erosion would result in eventual emergence of a core of pillow basalts and intrusions above sea level (Fig. 3.3b). These rocks have low porosity and permeability, and so limit inter- action between sea water and magma rising through them in dykes. Thus, once the intrusive core has emerged above sea level, a transition to effusive eruptions of lavas will take place and a subaerial shield volcano will form unconformably on the uplifted seamount-series rocks (Fig. 3.3c). Shallow- water hydrovolcanic rocks (hyaloclastites) and re-sedimented breccias will be preserved only on the flanks of the uplifted seamount.

Shield volcanoes and lateral collapses

Following the emergence of the first island shield volcano in central Fuerteventura, about 22 million years ago, two more shield volcanoes developed to the north and south during the following 10 million years, each persisting for perhaps 2–3 million years. The central volcano ceased its main phase of growth about 18 million years ago, but is in places capped by sequences of lavas that infilled valleys eroded into the older rocks. These rocks, of the Melindraga, Tamacite and Los Tableros forma- tions, range in age from 17.5 to 13 million years and are therefore of the same age as the shield volcanoes to the north and south. The oldest rocks of the volcano in Jandía are 20–19 million years old, but this volcano was mainly active 17–15 million years ago, whereas the volcano in the north was active mainly 14–12 million years ago. All three volcanoes are domi- nated by basaltic lava flows, with a few trachytic lava flows and near- surface intrusions. The dips of the lava flows are radial to three volcanic centres, marked by stars in Figure 3.2. Dykes intruded into the flanks of the volcanoes define volcanic rift zones that also converge upon these areas. Thus, each volcano appears to have been broadly similar to the major shield volcanoes of Tenerife and La Palma.

Within the Basal Complex in the Betancuria area, it seems likely that the youngest large intrusions – those that cut the major dyke swarm (trending north-northeast) rather than being cut by many dykes – are of the same age as the shield volcanoes. Among these the most notable is the Vega Ring Complex, just south of Betancuria itself (stop 3.3). It is likely to represent the roots of a summit volcano, perhaps comparable to Teide on Tenerife. Up-slope projection of the lava flows on the southern and eastern flanks of this volcano towards the Vega Ring Complex suggests that the volcano rose about 3 km above sea level.

It seems likely that, like the younger shield-stage volcanoes of the Canary Islands, the shield volcanoes of Fuerteventura underwent giant lateral collapses. As noted above, extensive **debris-avalanche deposits** have been mapped on the deep ocean floor off the western coast of the island (see Fig. 1.4). The curved western face of the Jandía Peninsula is highly suggestive of an ancient, deeply eroded collapse scar. It seems likely that these collapses were very large and that they removed the summits of the volcanoes. In the south, the projected position of the summit of the Jandía volcano lies well off shore today. Farther north, the deep levels of exposure in the Betancuria Massif may reflect the occurrence of a giant collapse, the scar of which extended as far east and north as the central valley of Fuerteventura, between Betancuria and the hills along the eastern coast of the island. However, given the very deep erosion that has affected the island since the Miocene, it is impossible to say whether it represents the product of a single giant landslide, with a volume in excess of $1000\,km^3$, or a series of smaller giant landslides. Evidence for the timing of this collapse or series of collapses is discussed further below (stop 3.5).

Post-erosional volcanism and sedimentation
Shield-stage volcanism on Fuerteventura ended about 12 million years ago, and was succeeded by a long period of erosion. Intermittent basaltic volcanism of the post-erosional phase of activity began again about 5 million years ago; but, in contrast to the very large volumes of post-erosional volcanism in Gran Canaria to the west, post-erosional volcanism in Fuerteventura has been comparable to that in Lanzarote, producing thin but laterally extensive alkali-basaltic lava fields erupted from groups of scoria cones. Some of these lava fields appear to be very young, because of the slow rate of erosion and weathering in the dry flat landscape of the island. However, although the youngest rocks are perhaps a few tens of thousands of years old, there have been no eruptions in the island in historical times. A few seismic swarms and an episode of steam emission near Pájara in 1915 are the only historical evidence that magmas may still be present at depth. Nevertheless, as in the case of Lanzarote, it is likely that Fuerteventura may yet experience eruptions in the future.

The recent geological history of Fuerteventura has been dominated by the effects of sea-level change. During periods of high sea level, most notably in the early Pliocene around 5 million years ago, when sea level was around 50 m above present, but also during more recent interglacial periods, shallow marine carbonate sediments and beach deposits have been emplaced. Similar carbonate sands have accumulated off shore on the wide shallow-water platform around the island. During sea-level **lowstands**, resulting from the growth of the great ice sheets, the platform emerged above sea level. Because glacial periods are also cold and arid, the carbonate sands were blown inland by strong winds (also a characteristic of glacial periods). There are extensive fields of white carbonate sand dunes in the north of Fuerteventura, around Corralejo, and they cover the whole of the neck of land between Jandía and the rest of the island. Earlier dunefields may have covered much of the island, but were dissolved away when warmer and more humid conditions returned; the evidence for their existence consists of the thick layers of re-precipitated carbonate or **caliche** that form hard crusts over large areas of Fuerteventura.

Logistics on Fuerteventura

It is just about possible to drive from one end of Fuerteventura to the other and return in a day, but the length of the island means that either a visit must be designed around two centres or else a base in the central part of the island is essential. Furthermore, Fuerteventura is less intensively developed for tourism than its neighbour to the north. The main resorts are at Corralejo, at the northern tip of the island, and around the Costa Calma and Morro Jable in the south. The capital, Puerto del Rosario, is primarily a dormitory for local residents. Perhaps the best base for a geological visit to Fuerteventura is one of the small resorts along the east coast, such as Caleta de Fuste, El Castillo or Gran Tarajal (Fig. 3.4).

The sparse population and the scale of the island mean that bus and taxi services are limited, so by far the best option for travel is a hire car. Roads are generally metalled and there is little need for four-wheel drive, although care is definitely required when driving on gravel roads in Jandía. Off-road driving in the sand dunes is strictly prohibited. The main areas of dunes, along with most of Jandía except for the populated south coast, the Macizo de Vega area of southeast Fuerteventura and the Vega de Río Palmas area around Betancuria, are protected regional natural parks; if you plan to sample in these areas, apply for a permit.[*]

Fuerteventura is one of the first of the Canary Islands to be covered by the excellent new series of 1:50 000 scale topographical maps, published by the Cartografica Militar de España. These are particularly useful because they carry the new FV-series road numbers.

Fuerteventura and Lanzarote are by far the driest of the Canary Islands, for the most part desert or semi-desert, with only the area around Betancuria being a little more humid. Thus, protection against sunburn and dehydration is essential: hat, sun cream and plenty of water (plus rehydration salts or isotonic drinks) are essential items. Tiempo sur ("southern weather"), when southeasterly winds bring hot, dry and dust-laden desert air direct from the Sahara, is especially trying. At other times, thin cloud layers can make conditions for photography highly variable.

---

[*] From the Department of Environment of the Canarian Government: www.gobcan.es/medioambiente

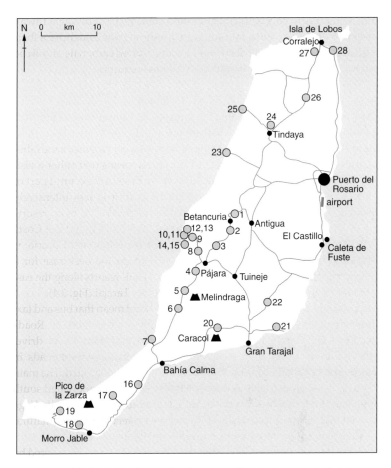

Figure 3.4 Map of locations on Fuerteventura, with main towns also indicated.

Seamount rocks and deep-ocean sediments

The main area of outcrop of the seamount series and deep-ocean sediments is on the western side of Fuerteventura in the Betancuria Massif, the largest and most humid area of high ground. A north–south road, the FV-30, runs through the area from Llanos de La Concepción through Betancuria to Pájara; a minor road between Pájara and La Pared provides access to the southern part of the area. A drive along this road, with short excursions to either side, provides an overview of the seamount series and its relationships to the younger rocks. However, perhaps the most interesting rocks are to be found in a small area around Puerto de La Peña, west

31

of Pájara, and a walking itinerary around these is described separately. These two itineraries will take a day each to follow in full. Smaller outcrops of Basal Complex rocks are also found on the coast farther north, and these are dealt with in the itinerary around northern Fuerteventura (pp. 51–55).

North–south traverse through the Betancuria Massif

This route is described beginning in Llanos de La Concepción, 17 km west of Puerto del Rosario on the FV-10 (turn off onto the FV-30 after 12 km) and 40 km south of Corralejo; shield-stage rocks seen en route are described separately (p. 51). From Llanos de La Concepción continue south through the smaller village of Valle de Santa Inés on the FV-30, following signs for Betancuria and Pájara. Beyond this village the road climbs up hill into the Betancuria Massif through a series of switchback bends.

Stop 3.1 Mirador Morro de Velosa   About 6 km from Llanos de La Concepción, a turning to the right is signposted to the Mirador Morro de Velosa, 1 km from the main road. This mirador provides an excellent view over the broad undulating valley that separates the Betancuria Massif from a range of hills to the north and east, known locally as cuchillos. These are distinctive: long narrow ridges, highest at the points closest to the massif and sloping down to the north and east, and separated by broad valleys with flat floors. They represent the eroded flanks of the shield volcanoes that originally capped the island, formed by lavas dipping down slope to the north and east. A radial drainage system developed on these slopes and, after the shield stage, the valleys gradually broadened until the areas of high ground were reduced to narrow ridges, the cuchillos.

The broad valley between the cuchillos and the Betancuria Massif conceals the contact between the latter and the shield-stage lavas under a cover of younger sediments and lavas. It was most probably eroded along a zone of soft rocks at the contact; the nature of these rocks is indicated by coastal exposures at the southern end of the massif (stop 3.7).

The rocks forming the hills around the mirador consist of rather altered submarine volcanic types, principally pillow basalts. The characteristic globular shapes of the lobes or pillows of these submarine lava flows can be identified in places. However, the most obvious structures in the rocks are the very abundant north-northeast-trending dykes, typically each less than a few metres thick but together forming more than 50 per cent of the volume of the sequence. In comparison, very few dykes cut the shield-stage lava flows to the north and east, leading to the conclusion that the majority of these dykes belong to the seamount stage of growth of the Fuerteventura volcanic edifice. The trend of the dykes is broadly parallel to the long axis of Fuerteventura (other than Jandía), suggesting that the

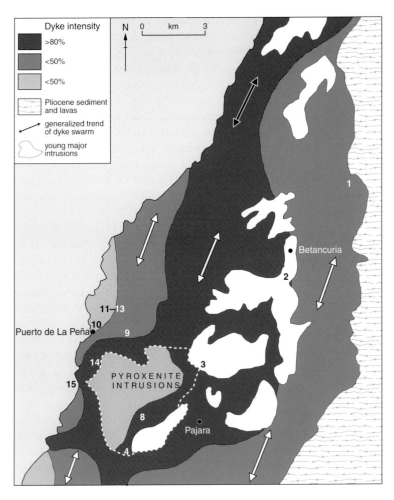

Figure 3.5 Distribution of dykes and major intrusions within the central part of the Betancuria Massif, with key stops indicated.

elongate shape of the volcanic edifice as a whole was controlled by the emplacement of these dykes and is therefore a feature that developed very early in its history. The overall distribution of the dykes is shown in Figure 3.5, along with the location of major intrusions that post-date some or all of the dykes and are considered at stops described below.

Continue south to the village of Betancuria, the first major Spanish settlement on the island, principally because the valley in which it is located is one of the few places in the island with sufficient water available for agriculture. It retains many examples of early colonial Spanish architecture.

Stop 3.2 Road south of Betancuria Exposures of rocks in and around Betancuria itself are limited, but roadcuts along the FV-30 1–2 km south of the village reveal an abrupt decline in the intensity of dykes: they form as little as a few per cent of the outcrops, which are instead dominated by coarser-grain intrusive rocks. These are principally highly porphyritic micro-**syenite**, with abundant alkali feldspar phenocrysts. They also contain many small fine-grain inclusions of darker rocks, richer in mafic minerals (principally pyroxene and amphibole) with characteristic lobate fine-grain margins. These **mafic enclaves** represent blobs of basaltic magma that were injected into the lower-temperature trachytic or phono-litic magma that eventually crystallized to form the syenites. The temper-ature difference between the two magmas resulted in rapid cooling and chilling of the mafic enclaves.

All of these rocks are much less heavily altered than the submarine volcanic rocks and the main dyke swarm. This indicates that they post-date the emergence of the island above sea level and the end of intense alteration by reaction with circulating heated sea water. This conclusion has been confirmed by radiometric dating. They are of the same age (about 20 million years old) as the main shield-stage sequence of the central Fuerteventura volcano. The co-existence of basaltic and felsic magmas indicated by the enclaves suggests that the volcano erupted rocks of both types from its central vents, although felsic lavas (trachytes and phono-lites) are relatively rare in the sequences in the cuchillos. There are no major ignimbrites on Fuerteventura comparable to those seen on Tenerife.

Stop 3.3 Degollada de Los Granadillos: the Vega Ring Complex Continue southwards on the FV-30 towards Pájara through a deep valley, the Bar-ranco de Palmas, with mountainsides on either side composed of green-weathering seamount-series rocks (mainly dykes of the main swarm) and paler-weathering major intrusions. Perhaps the most notable of these is best seen south of Vega de Rio Palmas, where the road climbs up a steep mountainside to Degollada de Los Granadillos, 3.5 km beyond Vega de Rio Palmas. A viewpoint, just before the road crosses the crest of a narrow arcuate ridge, offers the best place from which to view the Vega Ring Com-plex (Fig. 3.6). The arcuate ridge, which extends across the valley to the north and is cut by a narrow canyon, is formed by a ring of erosion-resist-ant pale purple-brown-weathering gabbro. This ring-shape (annular) intrusion is some 4 km across and at its centre is a circular intrusion of syenite that has been eroded away to form the valley in which the village is located. The two intrusions together form the Vega Ring Complex, one of the youngest major intrusions in the Betancuria Massif (note that they cut the main dyke swarm). It represents the feeder complex through which

Figure 3.6 View of the characteristically well jointed, pale-weathering gabbro of the Vega Ring Complex.

magma ascended to the summit of the central shield volcano about 20 million years ago.

The gabbro itself can be seen in roadcuts adjacent to the viewpoint (beware of traffic while examining them). It is a very coarse-grain pale crystalline rock, with many feldspar crystals up to 3–4 cm long. The rock does not correspond to any liquid magma in composition; rather, the feldspars crystallized out of a basaltic to trachybasaltic magma and separated from it to form a **cumulate** rock, leaving a melt of more felsic composition. The last stages of this process are evident in these rocks, which are cut by irregular veins and segregations of pink micro-syenite rock, containing feldspar and powdery altered crystals of nepheline. These represent the last melts to be squeezed out of the gabbro, which solidified before they could escape completely. Although most of the felsic rocks in the massif are discrete intrusions of syenite, formed from similar magmas produced at greater depths, these outcrops offer some insight into one of the mechanisms by which felsic rocks may be produced in oceanic-island volcanoes. The gabbros are also cut by a few fine-grain basaltic dykes; these later intrusions, emplaced along fractures in the gabbros after they solidified, are among the youngest rocks of all in the Betancuria Massif.

Continue south to the town of Pájara (a good place for lunch). In Pájara, turn left towards Puerto de La Peña, but after 2 km bear left at a fork on the road south to La Pared. This passes through hills formed by the north-northeast-trending dyke swarm. Between the basalt and **dolerite** dykes are thin slivers or screens of the country rocks, which in this part of the Basal Complex are not submarine volcanic rocks but coarse-grain intrusive rocks, but including pyroxene-rich rocks (pyroxenites) along the section of road running south for 2 km from the fork in the road.

Stop 3.4 Barranco de La Solapa   These rocks belong to the younger of two large pyroxenite intrusions, the larger and earlier of which is to the west in the Barranco de Pájara (stop 3.8). They are best examined at the south end of the pyroxenite intrusion, where the road crosses the Barranco de La Solapa about 2 km from the fork in the road. The pyroxenite contains many fewer dykes than the gabbros to the south, implying that they post-date much of the dyke swarm. Furthermore, immediately south of the pyroxenite, the gabbro and many of the older dykes have been metamorphosed by intrusion of the pyroxenite, producing compact granular hornfelses. The pyroxenite is considered to have formed by accumulation of crystals from a particularly hot and **primitive** magma, resulting also in the broad metamorphic aureole in the surrounding rocks.

Stop 3.5 Barranco de Fayagua: view of Melindraga   After about 8 km, the road passes over a low col into the broad northwest-trending Barranco de Fayagua. A parking bay on the right, where the road crosses the dry watercourse in the floor of the barranco, provides an excellent place from which to view the peak of Melindraga, at the head of the barranco to the southeast. The mountainside exposes a prominent unconformity (Fig. 3.7) between seamount-series rocks of the Basal Complex below and a sequence of subaerial, almost horizontal, lava flows above, which filled the palaeo-valley defined by the unconformity. The seamount rocks here are primarily dykes, resulting in the prominent stripes that run across the hillside almost parallel to the contours (but note the way that the dykes cut across gullies, indicating that they are almost vertical). The subaerial lava flows are basaltic and trachybasaltic, and they overlie a poorly exposed sequence of sediments derived by erosion of the Basal Complex rocks.

The age of the rocks forming the summit of Melindraga has been determined by radiometric dating: they are among the youngest rocks in the central shield volcano, at about 17 million years. Furthermore, the same flows unconformably overlie older shield-stage lavas farther to the southeast. The implication is therefore that the earlier shield-stage rocks and a substantial amount of Basal Complex rocks, with a thickness of 1–2 km, were removed from this area in the space of 3 million years. The most

Figure 3.7 The major unconformity between Basal Complex intrusive rocks (below) and the youngest shield-stage volcanic rocks on Melindraga (stop 3.5).

plausible explanation for this is that the western flank and summit of the central shield volcano were removed by one or more giant lateral collapses. Erosion of the floor of the collapse scar then produced a barranco system (as is the case with the Caldera de Taburiente on La Palma), on the floor of which these younger lavas accumulated.

Continue south along the same road for a further 3 km to a T-junction at the head of the long Barranco de Chilagua and park by this junction.

Stop 3.6 Barranco de Chilagua This barranco is also aligned along the north-northeast-trending main dyke swarm. In this area, as seen in road-cuts around this junction, the dykes are intruded into green, heavily altered and metamorphosed submarine volcanic rocks. It is difficult to identify the original nature of most of these, but they were principally basaltic pillow lavas with some hyaloclastite breccias. All rocks show evidence of metamorphic alteration at temperatures of around 300–400°C, producing green chlorite and amphiboles.

Continue down the Barranco de Chilagua. Roadcuts and small quarries by the side of the road expose more dykes intruded into pillow basalts and other submarine volcanic rocks, with a general southward decrease in the abundance of the dykes. The higher slopes of the hills on either side of the barranco are formed by shield-stage lavas unconformably overlying seamount volcanic rocks and the dyke swarm. The road emerges from the Betancuria Massif near La Pared; turn right for this small village and to the coast at Punta de Guadalupe, north of the beach and golf course.

Stop 3.7 Punta de Guadalupe Both Punta de Guadalupe itself, a prominent crag on the coast, and the coastal platform to the north, expose brown-weathering breccias. The rocks on the intertidal coastal platform are breccias with abundant clasts (1–10 cm across). These are mainly angular to sub-rounded **lithic clasts** of basalt, amygdaloidal and fresh to greenish grey (altered at low metamorphic grade), but red-oxidized basaltic scoria clasts are also present. The latter appear to have been reworked from sub-aerial scoria cones. The breccia on the platform grades upwards, in the low cliffs on the landward side and in Punta de Guadalupe, into a matrix rich breccia. As well as lithic clasts similar to those in the underlying rocks, this has a proportion of angular to ragged basaltic clasts with yellow altered glass rinds, and a matrix dominated by smaller clasts of yellow altered basaltic glass. This rock was produced by rapid erosion and re-working of a hyaloclastite, erupted at or near sea level in hydrovolcanic explosions driven by expansion of steam produced when sea water entered the volcanic vents.

The breccias are cut by several north-northeast-trending dykes 0.2–1 m

Figure 3.8   Reworked hyaloclastite breccias at Punta de Guadalupe (stop 3.7).

thick. These are irregular and segmented, with well developed chilled glassy margins reflecting emplacement in soft and water-saturated host rocks (Fig. 3.8). These rocks are among the few that record the transition from the submarine volcanic rocks of the seamount to the subaerial shield volcanoes. As noted in the introduction (see especially Fig. 3.3), this transition seems to have involved intense erosion and reworking of the hyaloclastites produced by hydrovolcanic explosions on the emergent island, until continued uplift over the growing intrusions in the core of the volcano elevated resistant intrusive rocks to sea level, where they formed a stable platform for eruption of the first shield-stage lavas. The rocks exposed at Punta de Guadalupe are on the flank of the seamount and represent sediments reworked onto the flanks by erosion of the hyaloclastites and other emergent-stage volcanic rocks. Similar rocks may underlie much of the central valley of Fuerteventura, between the Betancuria Massif and the shield-stage volcanic rocks forming the cuchillos to the east (stop 3.1), and are also exposed in the north of the island (stop 3.23).

La Pared is close to the resorts at the south end of the island; at the end of this tour it is possible either to stay in one of these resorts or to return to the east-coast resorts via the FV-2. Rocks visible from this road are considered in more detail below (pp. 48–51).

Puerto de La Peña area
The small village of Puerto de La Peña lies on the coast west of Pájara, in the centre of a small area that contains many of the most interesting rocks in the Basal Complex. The following stops are mainly linked by walks

north and south of Puerto de La Peña; note the precautions necessary for walking in the hot dry climate of Fuerteventura (pp. 17–18, p. 30). The itinerary begins at the junction of the Pájara–La Pared road with the road to Puerto de La Peña, 2 km west of Pájara.

Stop 3.8 Morro de Pedregullo   About 1 km north of the road junction, the route passes along the foot of the eastern side of Morro de Pedregullo. At 500 m before the road crosses the floor of the Barranco de Pájara on a bridge, it crosses the contact between the younger pyroxenite (stop 3.4) and an older but still primarily post-dyke-swarm pyroxenite intrusion. The contact is poorly exposed and the break in slope at the foot of Morro de Pedregullo marks the western margin of a zone of relatively young north-northeast-trending dykes associated with the younger pyroxenite intrusion. This dyke swarm has been preferentially eroded. To the west, the older intrusion itself has a sheeted structure, with alternating zones of pyroxenite, feldspar-bearing pyroxenite and gabbro. These features indicate that the extensional stress field that generated the north-northeast-trending dyke swarm (perpendicular to the trend of the dykes) persisted after emplacement of the main dyke swarm, perhaps to the beginning of the shield stage of volcanic activity.

Stop 3.9 Morro del Moral   The road continues north for 3 km across the broad area of low ground where the Barranco de Pájara joins the Barranco de Ajuí, and turns to the northwest and then west down the latter. The slopes of Morro del Moral to the south, and a series of roadcuts along the road itself, expose the dyke swarm trending north-northeast. The dykes are very abundant (50–80% of the total section) and many are intruded into or along the contacts of older dykes, but in places the original country rocks are exposed. These are metamorphosed and hydrothermally altered, but are still fine-grain dark mudstones and yellow siltstones. They belong to the lowermost unit (Basal Unit) of the Jurassic deep-sea sediment sequence exposed in this area, and were deposited on mid-ocean ridge basalts, 180 million years old (just as the Atlantic started to form or open), that occur in very small exposures in the hills to the south.

Continue to the coast at Playa de Ajuí, the small beach at Puerto de La Peña, and park there. Puerto de La Peña has a couple of small bars for a convenient break between walks to north and south of Playa de Ajuí. Here the walk to the north is described first, as it deals primarily with the Jurassic deep-marine sediments (continuing from stop 3.9), whereas the walk to the south deals primarily with seamount-series intrusions and associated metamorphism and deformation.

Stop 3.10 Playa de Ajuí The coastal cliffs on the north side of Playa de Ajuí are capped by flat-lying Pliocene marine sediments, but the lower parts of the cliffs near sea level are formed by much older rocks, cut by dykes of the north-northeast-trending swarm. These dykes form less than 50 per cent of the outcrop, and between them are yellow-weathering (green when fresh) metamorphosed limestones and mudstones of a distinctive pelagic bivalve-bearing unit (Fig. 3.9). Impressions of the bivalves have been preserved in the rocks despite the metamorphism. These are well bedded, dipping to the south at around 60°, but burrows and ripple

Figure 3.9 Steeply dipping, well bedded metamorphosed sediments cut by dykes at Playa de Ajuí.

structures in the sediments indicate that the rocks become younger north-wards, implying that the whole sedimentary sequence has been over-turned.

Stop 3.11 Caleta Negra   A path up the north side of Playa de Ajuí provides access to a coast path 40–50 m above sea level. This path runs along a bench cut into a coarse limestone, dominated by shell debris, with prominent cross bedding. Of early Pliocene age (about 5 million years old), this was deposited on a marine abrasion platform cut into the Basal Complex from the late Miocene to early Pliocene. It is in turn overlain by younger lava flows of the post-erosional volcanic activity, emplaced after sea level fell at the onset of glaciation later in the Pliocene. The top of the sedimentary sequence, about 50 m above sea level, is at the same height above present sea level as the worldwide Pliocene sea-level **highstand**. This implies that Fuerteventura has been completely stable over at least the past 5 million years. The underlying marine abrasion platform, and the striking uncon-formity between the Pliocene limestones and the Basal Complex, can be seen in the cliffs around Caleta Negra bay, 500 m to the north of Playa de Ajuí. The rocks below the unconformity include many dykes, which are intruded into Jurassic sediments, principally siltstones and mudstones. At the northern cliff of the Caleta Negra bay, the front of the post-erosional lavas can be seen forming pillows, as the flow front entered the Pliocene beach (Fig. 3.10).

Figure 3.10   Sequence of Pliocene rocks overlying unconformity above Basal Complex dykes and Jurassic sediments, Caleta Negra bay. The Pliocene rocks consist of shallow marine lime-stones overlain by dark pillow lavas.

Stop 3.12 Barranco de La Peña   About 1 km north of Puerto de La Peña, the coast path descends into Barranco de La Peña and continues up the barranco. Basal Complex rocks, below the Pliocene limestones and Recent lava flows, are exposed in the walls of the barranco. There are many north-northeast-trending dykes, intruded into a well bedded sequence consisting of interbedded mudstones and siltstones, the Main Clastic Unit of the Jurassic to early Cretaceous sedimentary sequence. The latter are **turbidites**, mainly composed of quartz, feldspar and other mineral grains derived from the adjacent continent. The turbidites indicate an increase of erosion on the continent and of sediment transport to the deep ocean floor. This and the older Mixed Clastic Unit indicate growth of a deep-marine sedimentary fan outwards from the Moroccan coast. Like the rocks to the south, the sequence dips southwards and is overturned.

The recent lava flows on the rim of the Barranco de La Peña to north and south were erupted from a vent about 5 km to the east. The flows descended into the barranco and filled it, and it has since been re-eroded as a deep narrow channel, re-exposing the older rocks in the lower walls.

Stop 3.13 Barranco de Los Sojames   After about 500 m, the path exits the Barranco de La Peña and continues northeastwards along the north side of a tributary barranco, Barranco de Los Sojames. The deep-marine sediment sequence is exposed in the floor and lower walls of this barranco and is again intruded by many dykes. The Main Clastic Unit is exposed in the lower 500 m of the barranco, and passes abruptly north into fine-grain pelagic limestone or chalk of early Cretaceous age.

At the top of the barranco, another 700 m farther northeast, the limestones are unconformably overlain by Tertiary volcaniclastic rocks, also heavily intruded by dykes of the north-northeast-trending swarm. These are primarily composed of basaltic clasts, but clasts of trachytic rocks are also present. Fossils found in this sequence are of Oligocene age. Unlike the Mesozoic sediments, these rocks are not overturned, indicating that the deformation of the rocks to the south was earlier. As discussed in the introduction to this chapter, the likely age of the deformation is mid-Cretaceous and its likely cause the regional-scale slumping that occurred all along the Atlantic continental margin of Africa at this time.

Return to Puerto de La Peña. Continue south of the village on the coast path; after about 200 m, bear inland along the Barranco del Aulagar.

Stop 3.14 Barranco del Aulagar   The rocks in the floor of this barranco, in the lowermost 1 km of its course, are Tertiary submarine basalts, heavily intruded by north-northeast-trending dykes. The low hills to the southeast belong to the major pyroxenite intrusion examined at stop 3.8 (see also

Figure 3.11 Segregation veins formed by deformation during partial melting in Basal Complex rocks close to a major intrusion, Barranco del Aulagar (stop 3.14).

Fig. 3.5). This has a well developed metamorphic aureole, the most spectacular expression of which is found about 1–1.5 km up the barranco. Here, within about 200 m of the contact of the pyroxenite to the east, the dark hornfelsed basaltic rocks have undergone partial melting and contain networks and arrays of centimetre-scale feldspar-rich veins (Fig. 3.11). These formed when the partially melted rocks underwent deformation, squeezing the melt out into the veins. Their abundance and orientation indicate shearing of the partially molten rocks parallel to the intrusion contact to the east.

Continue up the barranco for another 500 m to its head, and turn west to the coast path across an area of Pliocene calcareous sediments. A path leads down to the coast at Caleta de La Cruz.

Stop 3.15 Caleta de La Cruz The coastal rock platform at Caleta de La Cruz is cut in Basal Complex rocks. Many north-northeast-trending dykes are present, cutting rocks of the early Tierra Mala intrusion. These are mainly pyroxenites, with some gabbro and syenite veins. On the rock platform, at the north side of the bay, a 10 m-wide northwest-trending shear zone cuts these intrusive rocks and is itself cut by later dykes. The shear zone contains foliated fine-grain rocks, composed of pyroxene and **phlogopite**; this mica gives the rock its strong foliation. Also present are veins of carbonatite, mainly composed of calcite, which have a distinctive pitted appearance caused by dissolution by sea water. Reaction between the earlier pyroxenite and the hydrous carbonate melt represented by the carbonatite veins produced the pyroxene–phlogopite rock. Carbonatite veins occur through much of the Tierra Mala intrusion, representing one of the major occurrences of this unusual rock type in the Canary Islands.

The structures within the shear zone indicate right-lateral movement (far side moved to the right as you look across the shear zone). Similar northeast-trending shear zones are present elsewhere in the Tierra Mala intrusion. Together the two sets of shear zones were produced by broadly east–west extension, similar to the dyke swarms, but in a strike-slip stress field rather than an extensional stress field. This change in the stress field most probably reflects the effect of the emplacement of the major intrusions to the east.

Return to Puerto de La Peña along the coastal path.

Shield volcanoes in the south and east

This itinerary covers the main areas in the south and east of Fuerteventura, where shield-stage rocks are exposed. It begins at the roundabout on the FV-2 outside Bahía Calma on the Istmo de la Pared, between the main part of the island and the Jandía Peninsula. This is 66 km from Puerto del Rosario on the FV-2. The itinerary splits into two parts, Jandía and the east of the island and these can be followed separately, although it is assumed here that they are combined into a single day. If this is done, beware of fatigue, especially in view of the large amount of driving involved. Alternatively, if the tour is split into two days, with an overnight stay in the south of the island, time becomes available for walks in Jandía or on the beaches on either side of the peninsula (note: the west-coast beaches, Playa de Cofete and Playa de Barlovento, are not suitable for swimming because of an extremely dangerous undertow and longshore current).

Caution: in ordinary vehicles the gravel road beyond Morro Jable towards Playa de Cofete is passable only with care.

Jandía
Stop 3.16 Istmo de la Pared   The low isthmus between the main part of Fuerteventura and Jandía is mainly covered by a field of white sand dunes several kilometres across and nearly 100 m high in places. These can be viewed from various points along the 4 km stretch of the FV-2 south of

Bahía Calma. The dunes are composed of carbonate sand; as noted in the introduction, these originally accumulated as shallow-water marine carbonate sands off shore during sea-level highstands, and were then reworked by the wind onto the higher parts of the island during glacial periods when sea level was low. Particularly thick sequences accumulated in this area because it is in the lee of the Betancuria Massif with respect to the northeasterly trade winds. The sequences have been preserved as aeolian sands because this is one of the driest parts of the island, again because of its position in the lee of high ground. However, since sea level is now high, the sands are not being replenished and long-term erosion is a cause of concern.

Continue south on the FV-2 to the mouth of the Valle de Los Canarios, 10 km from Bahía Calma, and turn left up the valley.

Stop 3.17 Valle de Los Canarios Valle de Los Canarios, about 5 km long and more than 200 m deep in places, is the northernmost of a series of valleys, separated by ridges or cuchillos, arrayed in a radial pattern around the southern side of the Jandía Peninsula and converging upon a point off the northwest coast. Stop 3–4 km into the valley and view the valley walls. The upper part of the section is composed of southeast-dipping lava flows (mainly basaltic, with rare thicker and paler trachybasalt flows) that form the main part of the Jandía shield volcano. These flows are cut by very few dykes; they appear to have been erupted from vents up slope to the northwest. A few pale-weathering trachyte tuffs occur at the base of this sequence. The whole upper sequence has been radiometrically dated at 17–15 million years old.

A more varied sequence lies unconformably below these lavas and tuffs, in the lower walls of the valley. Dominated by lava flows at the top, lower down are scoria horizons and yellow hydrovolcanic breccias and ashes, recording explosive volcanism during the period when the Jandía volcano was emerging above sea level. It should be noted that the whole of this sequence is substantially younger (20–19 million years) than the age of emergence (22 million years) of the central shield volcano to the north, so it may owe its preservation to its position in the shelter of this older volcano. This lower sequence is cut by abundant east-northeast-trending dykes that may represent an early volcanic rift zone. These dykes are also truncated by the unconformity, suggesting that much of the section was removed in the 2 million years represented by the unconformity. It appears that Jandía may be composed of two distinct shield volcanoes, one on top of the other; a possible explanation for this is considered below (stop 3.19). However, for consistency with published work, the rocks are referred to as upper and lower sequences of a single volcano.

Return to the main road and continue southwest on the FV-2. The road winds around the southern end of the cuchillos to Morro Jable. Drive around Morro Jable on the main road, turning left on the all-weather gravel road to Cofete and Punta de Jandía, north of the port west of the town. About 2 km to the west, the road crosses the Gran Valle.

Stop 3.18 Gran Valle and Morro Mungia   The eastern side of the Gran Valle is dominated by southward-dipping lava flows of the upper part of the Jandía shield volcano. These are cut by many northwest-trending dykes, defining a dyke swarm that would have underlain a rift zone active during the later period of activity of the volcano. To the west, Morro Mungia is composed of lavas, scorias and hydrovolcanic breccias of the lower sequence of the volcano and these are cut by dykes with various orientations, but mainly north–south. These are aligned at about 120° to the east-northeast-trending dykes in the Valle de Los Canarios, suggesting that the shield volcano had a triple rift system similar to that seen in other Canarian shield volcanoes (see pp. 7–8). The implication is therefore that the extension that produced the north-northeast-trending dyke swarm in the Basal Complex disappeared after about 20 million years ago.

Farther west along the road, the unconformity between the lower and upper sequences of the Jandía volcano is visible on the western side of Morro Mungia. Continue 5 km westwards along the northern side of a broad platform 30–50 m above sea level. This is covered by alluvium worked down slope from the mountains, but is principally formed by Pliocene marine sediments over an erosion surface cut into the shield volcano, whose rocks are exposed in the cuchillos to the north.

At a road junction at the mouth of a broad valley (Barranco de Agua Ovejas) turn left onto a gravel road winding up hill to the north. Most rocks in this area belong to the upper sequence of the Jandía shield volcano. Here are dykes trending both northwest (similar to those in the east wall of the Gran Valle) and northeast. The latter are on the margin of a well developed dyke swarm in the hills to the west, indicating that multiple rift systems were present in the upper sequence of the shield volcano as well.

Stop 3.19 Degollada de Agua Ovejas   The pass at the top of the barranco, the Degollada de Agua Ovejas, has a car-park, from above which it is possible to view the whole of the northern side of Jandía (Fig. 3.12). This is very different from the southern side, being a concave arcuate slope with a distinct resemblance to the younger collapse scars of the western Canary Islands, especially, El Golfo and El Julan in El Hierro. Although it is very much older than either and has been substantially modified by erosion, the occurrence of landslide deposits off shore (Fig. 1.4), and the presence of a substantial embayment in the submarine contours to the north, confirm that it is also a collapse structure.

The cliffs on either side of the pass and the headwall of the embayment to the east indicate that the lavas in the cliffs dip southwards and eastwards (into the cliffs) throughout. This, together with the pattern of convergence of the dykes, indicates that the summit of the Jandía volcano lay to the north and has been removed by the formation of the embayment. Reconstruction of the volcano indicates that the summit did in fact lie some 3 km off shore, north of the small village in the embayment, the Casas de Caleta around 3 km east of this viewpoint. Retrace the route to Bahía Calma on the FV-2.

Eastern Fuerteventura

As noted earlier, the centre of Fuerteventura is occupied by a broad lowland area, the Central Depression. This is about 100 m above sea level and is filled by sediments from the hills on either side and by post-erosional volcanic rocks. These rocks obscure much of the contact between the Basal Complex and the oldest part of the shield volcanoes (see stops 3.7 and 3.23), but there are sites of interest in the area.

Figure 3.12   The north coast of Jandía from Degollada de Agua Ovejas (stop 3.19): Betancuria Massif on horizon.

From Bahía Calma, drive towards the north and east on the FV-2. The road initially runs along the coast at the downslope ends of cuchillos running south from the Basal Complex. At Playa de Tarajalejo, 12 km from Bahía Calma, the road turns inland and 4 km to the north enters the Central Depression.

Stop 3.20 Caracol   About 3 km farther on, stop to view the face of Caracol, the 467 m peak to the south. This peak exposes much of the thickness of the central shield volcano in this area. This has been divided into three units, but the contact between the lowermost and the upper two is most evident, being marked by an unconformity, with some topography on it, overlain by several tens of metres of coarse unbedded sediments. Although there is a marked reduction in dip angle of the lavas above the unconformity, relative to those below, there is no significant gap in the ages of the rocks above and below the unconformity. This suggests that the sediments record at most a brief hiatus in volcanic activity or may represent deposits from (or reworked from) a volcanic debris flow. Relatively few dykes are present in the whole of the shield-volcano sequence.

On the north side of the road at and to the east of this locality are two low hills, Montaña del Moro and Montañeta de Agando. These are the principal outcrops of Basal Complex rocks (principally submarine basalts

with dykes) in this area; the contact with the shield volcano is in the valley along which the road runs and is poorly exposed.

Continue 7 km eastwards to the junction of the FV-4, and turn south on this towards Gran Tarajal, the second town of the island. 1 km outside Gran Tarajal, turn left onto the road to Faro de la Entallada at the south-eastern corner of the island (after 4 km, turn left again outside the village of Las Playitas). After a few kilometres, the road enters the Parque Natural de los Cuchillos de Vigan and climbs up the steep side of the Cuchillo de Entallada, on the crest of which the lighthouse ("faro") is located.

Stop 3.21 Faro de la Entallada The Faro de la Entallada is one of the very few places where a well defined route exists to the top of one of the cuchillos. It offers excellent views of the central shield-volcano rocks in this area. These mainly belong to the two older sequences of this volcano: the unconformity and intervening sediments are visible in places. To the west, along the southeastern coast, the rocks dip southwards, whereas to the north they dip eastwards. The change in dip direction is accompanied by an abundance of southeast-trending dykes in this area. This suggests that a volcanic rift zone ran southeast from the summit of the central shield volcano in the Betancuria Massif (stops 3.2 and 3.3) and that eruptions from it were sufficiently frequent and voluminous to build up a sloping topographic ridge comparable to those associated with rift zones in the western islands (e.g. the Cumbre Vieja in La Palma); the cuchillos to the west represent the eroded remnants of the southern flank of this ridge, and those to the north the remnants of the eastern flank.

Return to the FV-2 and continue northwards across the flat landscape of the Central Depression. About 3 km north of the junction, the road begins to skirt the edge of the Malpaís Grande, the field of young post-erosional lavas that fills the eastern part of the Central Depression.

Stop 3.22 Caldera de los Arrabales About 3 km farther north again, the road passes the Caldera de los Arrabales, the southernmost of a north–south-trending line of scoria cones spread over a distance of 10 km that fed the lava flows of the Malpaís Grande. Arrabales and the other cones, visible in the distance to the north, rise up to 200 m above the level of the Central Depression, and the total area of the volcanic field is some 30 km^2.

About 6 km to the north the road crosses a flow from the Malpaís Grande that escaped from the Central Depression and flowed to the sea at the east coast of the island, and then turns northwestwards up a valley between two of the most visually impressive cuchillos, Morro de la Pared to the west and Agudo to the east. The latter shows with exceptional clarity the characteristic long sloping ridge of a cuchillo, which slopes down to the east along with the dip of the shield-stage lavas that form the cuchillo.

At the northern end of this valley, the FV-2 turns eastwards to the coast and should be followed if you are staying in Puerto del Rosario or one of the east-coast resorts; those staying in the extreme north may wish to continue northwestwards and on to Corralejo or adjacent places via the more interesting landscape around Antigua and La Oliva.

Northern Fuerteventura
Erosion, sedimentation and post-erosional volcanism

Northern Fuerteventura was the site of the third of the major volcanoes that made up the shield-stage activity of the island, but, like those in the centre and south, this was deeply eroded. In many places along the west coast, post-erosional lavas and sediments of Pliocene to Recent age unconformably overlie submarine volcanic rocks and intrusions of the seamount stage of activity. This is a continuation of the pattern of asymmetry seen farther south where the Betancuria Massif exposes Basal Complex rocks at the west coast of the island (Fig. 3.2) and suggesting that the line of collapse structures along this coast may have extended to close to the northern end of present-day Fuerteventura.

This itinerary begins at the junction of the FV-20 and the ring road (FV-3) around Puerto del Rosario, and ends at Puerto del Rosario after passing through Corralejo; if staying in the latter place it may be more convenient to visit stop 3.28 first, on your way south to Puerto del Rosario.

Drive westwards on the FV-20 for 12 km, through a broad valley with cuchillos formed by remnants of the northern shield volcano on either side (this is the same route followed to reach the start of the tour through the Betancuria Massif, p. 32). Shortly after passing through the village of Casillas del Angel, turn right onto the FV-30, but at a crossroads 2.5 km farther on turn right (north) onto a minor road signposted to Tindaya. This leads up the eastern side of a broad valley that trends northwards from the Central Depression. After 3.5 km, fork left onto the road to Puertito de Los Molinos. The low hills to the north of this road are of Basal Complex (seamount series) rocks, but also mark the approximate position of the centre of the northern shield volcano, as deduced from the convergence of dykes and lava-flow slopes in the cuchillos upon this area.

Stop 3.23 Puertito de Los Molinos   Park in the village and proceed south along the coast. The rock platform along the shore, a few metres above the present sea level (which you may prefer to view from the top of the slope behind it), exposes many dykes cutting submarine volcanic breccias. The

51

older dykes all trend northeastwards but are notably inclined, tending to dip to the northwest rather than being vertical. This suggests that they may have been tilted after emplacement. The younger dykes, in contrast, are more nearly vertical than the dykes they cut and they commonly strike more east–west. They do not show the greenish hydrothermal alteration of the older dykes and they are likely to have been emplaced after emergence of the volcano above sea level.

The breccias between the dykes contain large fragments of pillow basalt and hyaloclastite, suggesting that they were emplaced in shallow water and are among the youngest rocks of the seamount series or belong to the mostly missing stage of emergent volcanism (see pp. 26–28). This interpretation is supported by the occurrence in breccias along this section of fragments of coral and algal reef, subaerial lavas and also clasts of dolerite, gabbro and pyroxenite. This suggests that the breccias were deposited on the submarine slopes of the emerging island at the stage when uplift and erosion had exposed the intrusive core of the rising seamount and the island was beginning to establish itself as a permanent feature about 21 million years ago (Fig. 3.3). The occurrence of reefs around the island is somewhat surprising in view of the continuing volcanic activity implied by the other rock types, but it will be noted that, in contrast to the corresponding outcrops at Punta de Guadalupe at the southern end of the Betancuria Massif (stop 3.7), this area is at the windward end of Fuerteventura and so the sea would be kept relatively clear of ash and debris falling from the hydrovolcanic eruption columns.

Return inland and drive north to the junction with the FV-10, then north for 2 km on this road before turning left into Tindaya. The road passes around the north side of the village; turn left onto a minor road leading to Montaña de Tindaya.

Stop 3.24 Montaña de Tindaya   Montaña de Tindaya, an isolated, distinctively jointed and pale-weathering crag (Fig. 3.13), was formed by an intrusion of trachyte, one of the few occurrences of felsic rock in the northern shield volcano. In the past it has been quarried as an ornamental stone, but it is now protected as an archaeologically significant site. Like many isolated peaks and pinnacles in the Canaries (e.g. Roque Nublo on Gran Canaria) it was a Guanche religious site. The energetic and agile may wish to follow a path that leads up the steep southern crest of Tindaya to the summit.

Return to the road past the village of Tindaya and continue westwards to the coast across a broad flat plain around 50 m above sea level. This is the Pliocene marine platform, mainly covered by relatively old post-erosional lava flows and alluvial sediments from the hills to the east.

Figure 3.13 View of Montana de Tindaya from the southeast.

Stop 3.25 Punta de Paso Chico Park at the end of the road and walk south-wards along the coast platform. The platform is covered by a 20–30 m-thick sequence consisting of a late Pliocene post-erosional lava flow overlying white to pale-yellow limestone of early Pliocene age. This in turn overlies an impressive planar unconformity, several metres above present sea level, below which are Basal Complex rocks: dark-green hydrothermally altered submarine volcanic rocks and abundant dykes of the seamount series. The whole of the northern shield-volcano sequence, which is inferred to have been more than 1 km thick in this area close to its central summit, was removed between the end of its activity about 12 million years ago and the erosion of this marine abrasion platform before the early Pliocene, about 5 million years ago. It seems very plausible that this in-volved lateral collapses, to produce the landslide deposits on the ocean floor to the west (see Fig. 1.4).

Return to Tindaya and drive north on the FV-10. After 6 km the land-scape changes from the low eroded hills of the northern shield volcano to a flat plain punctuated by scoria cones and covered by black lava flows and yellow-to-white aeolian sands and caliches. The main road turns left on the outskirts of the village of La Oliva, but you should continue into La Oliva, through the village towards Corralejo and past the southeastern side of the Monumento Natural del Malpaís de La Arena.

Stop 3.26 Malpaís de La Arena Stop after 6 km to view the large scoria cone of La Arena ("arena" meaning "sand", also used locally as a name for

53

lapilli) from the east. The intervening ground is occupied by yellow caliche derived by dissolution and re-precipitation of carbonate from wind-blown sands; a few old fossil dunes remain. These old aeolian deposits are overlain by the black lavas erupted from La Arena, visible in the distance.

Continue northwards towards Corralejo. After a further 4 km, the road crosses onto an alkali basaltic lava field. This large area of lava flows covers around 50 km² of northern Fuerteventura and most of it appears to have been erupted from a northeast-trending line of overlapping scoria cones that form a ridge up to 270 m high to the west of the road. These are being actively quarried for road and building stone, so tracks and paths in this area are subject to change. In general, access is easiest from a roundabout 2 km south of Corralejo where the road joins the FV-1. A gravel track and paths lead west from this roundabout into the Montañetas del Morro Francisco at the northern end of the chain of scoria cones.

Stop 3.27 Montañetas del Morro Francisco and Bayuyo   This area of very rough ground is formed by lava that emerged from the breached scoria cone of Bayuyo, around 1 km to the southwest (Fig. 3.14). All the cones in the chain are breached in one direction or another, and, through these breaches, lava flows emerged to produce the large flow field. In general terms, the eruption that produced these scoria cones and lavas was comparable to the famous 1730–36 eruption on Lanzarote in style of activity if not quite in scale, although, as it is prehistoric, its duration is not known. However, the lack of hydrovolcanic activity compared to Lanzarote is notable and suggests that these eruptions took place when sea level (and therefore the local water table) was low.

If time and the progress of quarrying permit, the best route to Bayuyo is via a track that leads to the northern edge of the Montañetas del Morro

Figure 3.14   The breached scoria cone of Bayuyo, viewed from sea to the northeast; the town of Corralejo extends along the coast below the scoria cone.

Francisco, which then gives access to a path along the northwestern side of the chain of scoria cones; a stiff scramble up the flank of Bayuyo leads to the summit. The whole walk from the roundabout is about 4 km and at least two hours should be allowed for the rough ground, but the view from the summit over northern Fuerteventura and across the strait to Lanzarote is spectacular.

Return to the roundabout. Corralejo is best avoided unless you are staying there or intend taking the ferry to Lanzarote or Isla de Lobos.

Stop 3.28 Arenas de Corralejo Turn east along the FV-1 at the roundabout. The road passes onto white carbonate sands that overlie the Bayuyo lavas, indicating that these pre-date the last major phase of active dune migration. The road continues through the dunes for the next 8 km, but the best places to stop are in the first few kilometres, as the dune tops provide views to Isla de Lobos, Bayuyo and Lanzarote.

The dune sands are composed of reworked marine carbonate organic sands that accumulated on the shallow sea floor between Lanzarote and Fuerteventura during previous interglacial periods when the climate was subtropical, as at present in the Gulf of Guinea, and were then reworked south and west by the prevailing trade winds during the most recent glaciation when sea level was low; winds were generally stronger and the climate of Fuerteventura was extremely arid. This accounts both for the composition of the sands (volcanic grains are extremely rare) and for the fact that they occur on the windward side of the island, with no apparent source up wind. An important corollary of this is that the sands are essentially a non-renewable resource, hence the importance attached to prevention of further erosion of the dunes by human activity.

Continue south along the coast road to Puerto del Rosario.

Isla de Lobos ("island of the seals")
A visit to Isla de Lobos takes most of a day, since the ferry leaves Corralejo (from the port at the northern end of the town; allow plenty of time to get through its congested streets) at 10.00 h and returns at 16.00 h. The island is uninhabited but is a pleasant place for a lazy day spent mostly on the beach; it takes about two hours to walk around. It consists of alkali basaltic lavas erupted from the partly eroded Caldera de Lobos on the northwestern corner of the island. This lies directly in line with the Bayuyo line of scoria cones, suggesting that it may have been formed in the same eruption: in this case the line of vents would have been at least 10 km long.

Chapter 4

Lanzarote

Lanzarote is the most easterly island of the Canaries, only 140 km from the African coast. Elongated in a northeast–southwest trend, parallel to the continental margin, it is 60 km long and 20 km wide, an area of 862 km^2 (905 km^2 including the small northern islets of La Graciosa, Montaña Clara and Alegranza). The topography is characteristic of mature islands, as is evident when comparing the shaded-relief images of Lanzarote (Fig. 4.1) and La Palma (Fig. 8.1). The landscape is dominated by deeply eroded volcanoes, U-shape barrancos and precipitous cliffs, separated by wide lowlands covered with organic aeolian sands and recent volcanoes.

From a geological point of view Lanzarote is not an island but a northern prolongation of Fuerteventura. The two islands are separated by the narrow and shallow (40 m maximum depth) stretch of sea, La Bocaina, and so form a single volcanic edifice, the two halves of which are connected above water when sea level is low during glacial periods. Both islands are in a very advanced (post-erosional) stage of development. Only the lack of significant subsidence (Ch. 3) means that these islands are still emergent, although they have lost most of their original subaerial bulk as a result of catastrophic mass wasting and erosion.

Volcanic activity in Lanzarote extends from the mid-Miocene to the present day. The island was formed by two independent shield volcanoes: the southern Ajaches volcano (now reduced to 560 m high) and the northern Famara volcano (now 670 m high). A central plain (less than 200 m above sea level), covered by recent volcanoes and lava fields, connects both massifs to form the present-day island of Lanzarote. This recent activity has covered much of the flat surface of the island in extensive but very thin volcanic fields, through which older rocks protrude. Because of the area covered and the slow colonization of these young lavas in the dry climate, the island appears to be more active than it actually is, especially when the age data on the youngest eruptions discussed below are taken into account. In particular, the rocks of the 1730–36 eruption mean that Lanzarote, the "island of a thousand volcanoes", is visited mainly because of these features, whereas La Palma, in the juvenile stage of shield building

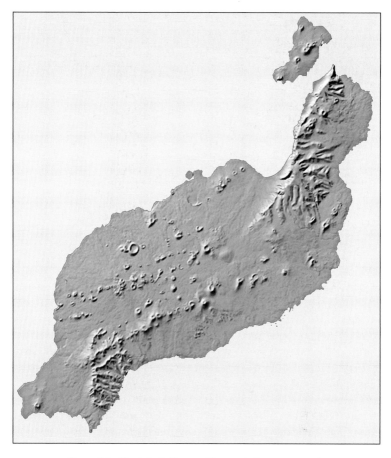

Figure 4.1 Shaded relief image of Lanzarote (image GRAFCAN).

and much more abundant recent volcanism, attracts visitors mainly because of the forests and erosive features (calderas and barrancos) that have developed on its more humid and steeper slopes (Ch. 8).

The modern geological study of the Canaries began in Lanzarote, but this led to confusion as "Old" and "Recent" series of rocks were defined in each island (Table 4.1), even though many rocks of the "Recent series" of Lanzarote or Fuerteventura are much older than the "Old series" of La Palma or El Hierro. It is therefore much more useful to consider the Canary Islands in terms of the characteristic stages of development of volcanic oceanic islands (shields and post-erosional stages), as explained in Chapter 1. Lanzarote is an excellent example of a post-erosive island, where both stages of growth are distinct and very marked.

Table 4.1 Different terminologies in the stratigraphy of Lanzarote.

Fúster et al. (1968) Basaltic Series	Carracedo et al. (1998)	Unit
IV 1–5 m beach*		Holocene and historical volcanism
III 10 m beach	2. Post-erosive or rejuvenation stage	Fissure volcanism
II 50 m beach		Peripheral volcanism
I	1. Shield-building stage	Famara volcano (10.2–3.8 million years) Los Ajaches volcano (14.5–5.7 million years)

* Raised or fossil beaches

The Ajaches shield volcano

This old shield volcano is located at the south end of the island. With a diameter of about 14 km, 608 m high and intensely eroded, it is only a fraction of the original volcano that extended originally at least from the Salinas de Janubio to the west to the area of the airport to the east. A westward-directed giant collapse and subsequent marine erosion have removed the northwest and central parts of the volcano. Marine erosion carved extensive abrasion platforms that have since been covered with recent volcanics (Fig. 4.2).

The remains of Los Ajaches volcano consist of a monotonous sequence of relatively thin basaltic lava flows, dipping to the south and southeast. Dykes are abundant in the northern sections, near Femés. Massive and deeply weathered trachytic lavas outcrop at the southwest coast (Janubio) and the south coast near Papagayo. These lavas may correspond to late-stage **differentiated** eruptions of the Ajaches shield volcano, similar to those described in the Taburiente and El Golfo volcanoes in La Palma and El Hierro. The lavas are covered by a thick carbonate crust (calcrete or caliche) and, in places, are visible only through erosive windows in the calcrete. The relief is characteristic of very old mature volcanoes: smooth interfluves and U-shape barrancos, partially filled with thick alluvial deposits.

Radiometric ages from the Ajaches shield lavas range from 19 to about 6 million years, but magnetic-stratigraphy studies reduce this period, suggesting that the main shield-building stage may extend from about 14.5 to 13.5 million years ago, with post-erosive volcanism extending to 5.7 million years ago. Most of the lavas from the Ajaches volcano are basanites and alkali basalts, except the trachytes mentioned above.

The Famara shield

The Famara volcano forms a 24 km-long, northeast–southwest-trending elongated shield in the northern part of the island. The western half has been completely removed, leaving a spectacular 23 km-long and 600 m-high cliff. This most probably represents the eroded scarp of an old giant gravitational collapse, although seaward marine erosion alone may have been sufficient to produce this scarp after near 4 million years without important eruptive activity.

Two main phases can be defined in the development of the Famara shield: the main shield-building stage, from about 10.2 to 8.7 million years

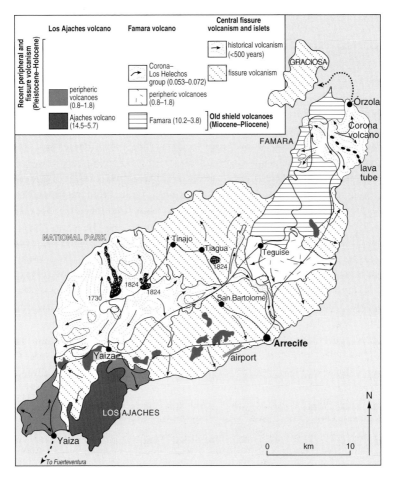

Figure 4.2 Simplified geological map of Lanzarote.

ago, and post-erosional volcanism from about 6.5 to 3.8 million years ago, with some recent residual rejuvenation activity. Most of the Famara volcano lavas are basanites, with only a few alkali basalts.

As in the Ajaches shield, the topography is mature. The dissected plateau at the top of the present massif – the eastern flank of the old shield – shows clear evidence of inverted relief: the present interfluves are the former barrancos filled with lavas, whereas the present barrancos have been truncated by retreat of the cliff. However, the Famara cliff is presently inactive and draped with continuous screes that have coalesced to form a colluvial coastal platform. The southeast flank of the volcano, formed by lava flows dipping east and southeast, have smooth interfluves and U-shape barrancos, with thick scree and alluvial deposits filling the barranco floors.

Post-erosional volcanism

Recent volcanism of the post-erosional stage of activity is represented by extensive but thin **Pleistocene** lava fields and chains of volcanic cones orientated in a general southwest–northeast trend. Holocene eruptions are very few, probably limited to the historic 1730–36 and 1824 events. The Corona and Los Helechos Group on the northern Famara shield, previously believed to be Holocene, have been dated at more than 50 000 years old, indicating that post-erosional eruptions on Lanzarote are much less frequent than was previously thought. As in Gran Canaria (Ch. 5) the volume of post-erosional rocks on Lanzarote is greater than in the Hawaiian Islands, but the style of volcanic activity, and in particular the frequent occurrence of hydrovolcanic explosions to produce tuff rings, is in many ways comparable to that of the "type" post-erosional volcanic sequence, the Honolulu volcanic series on the island of Oahu, which includes the famous Diamond Head tuff ring. The reasons for this are discussed below.

The ages of young volcanoes on Lanzarote
The belief that the Corona volcanic group (Los Helechos–Montaña Quemada–Corona) is very recent, less than about 3000–5000 years old, was based on the good preservation of these volcanic features. However, these volcanoes have been dated by Hervé Guillou (CEA–CNRS, France; personal communication) at 53 000 years old (the Corona volcano) and 72 000 years old (the Los Helechos volcano).

The Corona volcano produced one of the largest known lava tubes (7.5 km long, with some sections reaching 35 m in diameter), large enough to house a restaurant and a theatre. The last 1.6 km of the lava tube is submerged to a depth of 80 m, preserving sections up to 44 m wide at the end of the tube. The apparent formation of the lava tube under water is discussed in stop 4.15.

Historical eruptions

The eruption of 1730 to 1736 is the third largest basaltic fissure eruption to be recorded in historical times (after the 934 Eldgja and 1783 Laki eruptions in Iceland); no comparable eruption has lasted longer except for the continuing Pu'u O'u eruption on Kilauea in Hawaii (19 years at the time of writing). It was considerably longer (68 months), compared with less than three months for the second largest historical eruption in the Canaries, and larger (volume of about 700 million m^3), compared with 66 million m^3 for the second largest historical eruption in the Canaries. Over 30 volcanic vents were formed in five main multi-event eruptive phases (Table 4.2), aligned in a 14 km-long, N80°E-trending fissure that fed a lava field that covers much of the central plain of Lanzarote and also extended the island into the sea along much of its western coast. Accurate reconstruction of this eruption was greatly facilitated by eye-witness accounts, such as the report of the parish priest of Yaiza (included in the work of von Buch 1825), and particularly the official reports of the local authorities to the Royal Court of Justice, found on file by Carracedo & Badiola in 1989 in the Spanish National Archives in Simancas (Carracedo et al. 1990).

In 1824, a short eruption, with three vents aligned along a N70°E, 13 km-long fissure (stop 4.25), produced small amounts of basaltic lavas.

Logistics on Lanzarote

The extensive beaches developed during the post-erosional phase and the above-mentioned successful marketing of Lanzarote as "the island of a thousand volcanoes" attract more than 1.7 million visitors annually. The impression is of a rather hasty and artificial development. Fortunately, tourist resorts are concentrated on the geologically dull southern coast, leaving the more spectacular remainder of the island relatively untouched. There are many hotels and holiday apartments in a wide price range in Arrecife, Costa Teguise or Puerto del Carmen, in addition to small pensiones or apartments in villages such as Orzola in the north or Playa Blanca in the south.

The inexpensive hire cars are the recommended form of transport, four-wheel drive being ideal for exploring tracks. Lanzarote is a relatively

Table 4.2 Main eruptive phases in the 1730 eruption of Lanzarote. (Carracedo et al. 1992).

Phase	Date and duration	Eruptive vents	Accounts		Main features
5	April 1736 (end of eruption 16 April) Mid-March to beginning of April 1736	Montaña Colorada Montaña de Las Nueces			Eruption migrates east to the initial area of the eruption. Tholeiitic, very fluid lavas reach the coast 23 km afar. Abundant olivine, gabbro and quartzite xenoliths.
4	?	Montañas del Fuego			Very long stationary period of high effusion-rate episodes forming many overlapping volcanoes at the centre of the main fissure (Montañas del Fuego). Basaltic lavas.
3	Dec. 1731 to Jan. 1732? July 1731? June 1731?	Calderas Quemadas Caldera Raja Volcán El Quemado	Parish priest of Yaiza	Real Audiencia de Canarias	Abrupt jump of the eruption 12 km to the west and into the sea. Eruption follows a W–E constant progression.
2	March–June 1731	Montañas del Señalo			Opening of the main N80°E fissure. Lavas more silica-saturated and with few xenoliths.
1	10 Oct. 1730 to 1 Jan. 1731 10–31 Oct. 1730 1–13 Sept. 1730	Pico Partido Caldera Santa Catalina Caldera de Los Cuervos*			Eruptive vents aligned in a NW–SE fissure, still not related to the main N80°E main fissure. Lavas very SiO_2 undersaturated and olivine xenolith rich (up to 30–40% of the lavas).

* Initial vent of the 1730 eruption.

small (some 60 km from one end to the other) and predominantly flat island that can easily be seen by car in a few hours, using the many roads and tracks. The only dangerous areas are the highly unstable tracks along the rim of the Famara cliff. For safety, always use the designated vantage points described in stops 4.6–4.12. Never leave personal belongings in parked vehicles.

For those planning to visit the islet of La Graciosa – the only one not requiring permission – an excellent inexpensive ferry plies the 20-minute route several times daily from Órzola.

The National Park of Timanfaya covers most of the 1730–36 and 1824 eruptions and is one of the most interesting areas in Lanzarote, but there are many restrictions. Walks are guided and by appointment only. Book months in advance (tel. 34-928-840839). Rock or plant sampling is permitted for scientific purposes only and official permission is required from the park authority. Private vehicles must be left in the car-park, and the journey continued in the coach provided. The perimeter of the park was defined in 1974 before the extent of the historical eruptions was fully appreciated; consequently, several of the initial emission centres of the 1730–36 and 1824 eruptions described in the eye-witness accounts were not included. These and other areas of interest related to the eruptions can be explored with few restrictions, as described in detail on pp. 81–94.

The most interesting of the many restaurants in Lanzarote are those farthest from the tourist areas. Although both Lanzarote and Fuerteventura are drier and more arid than many deserts, with annual rainfall around 150–200 mm, most of the year the moisture-laden trade winds blow from the northeast. For long walks, particularly in summer or in "tiempo sur" (hot dry winds from Africa), prepare a snack that includes salted nuts and drink plenty of water (alternatively, take isotonic beverages). Do not be misled by the large areas devoted to horticulture. They are the result of an ingenious technique developed after the 1730–36 eruption, as is discussed later (p. 83). In recent years, a much more productive "crop" is being cultivated: mass tourism. The almost two million tourists who visit annually "dilute" the local population (85 000) and the formerly strong character of Lanzarote. However, many tourists visit only on organized one-day trips from the major islands and are taken on a specific route that includes the official national park, Los Jameos del Agua, etc.). With the exception of such places, one can still walk most of the island and seldom meet another human being.

Despite the desert-like landscape, brilliant sunshine is not always guaranteed. Unlike the islands with higher relief, where the trade winds bring rainfall and form the "sea of clouds" on the northern seaward flanks, leaving the south dry and sunny, the trade winds pass unchecked over the older lower islands, frequently generating a low-lying cloud cover that is very frustrating for the amateur photographer. Nevertheless, Lanzarote is very photogenic. There is little vegetation, with the exception of a few palm trees that, together with the geometric profiles of the volcanic cones, contribute to the spectacular scenery. Bring as many rolls of film as possible and take advantage of the early morning. Excellent results are obtained at sunrise, whereas sunsets are best photographed from Femés. Places offering magnificent views are noted in the itineraries section, although access may be somewhat difficult in certain cases.

The proposed itineraries are shown in Figure 4.3.

The Los Ajaches Massif

Stop 4.1 Erosional remnants of the Los Ajaches shield volcano From Arrecife (or the airport) take the road to Yaiza (LZ-702). Past an industrial area with some eroded basaltic cinder cones of the central recent fissure eruptions, the road cuts through several smoothly rounded hills. These are the eroded remnants of the old Los Ajaches shield volcano. The lavas that

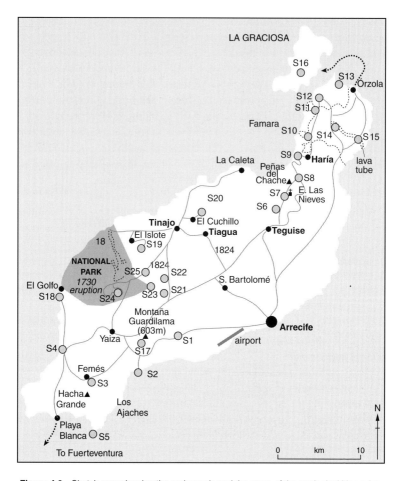

Figure 4.3 Sketch map showing the main roads and the stops of the geological itineraries.

form these hills are thick weathered basaltic flows capped by recent volcanics or with a caliche crust and aeolian sands. The structure of these erosional remnants of the old shield can be observed in detail in the quarries exploited for building and road-making materials.

Stop 4.2 Playa Quemada   Past km-16 turn left to the village of Playa Quemada, 4 km distant. The basaltic flows that form the Los Ajaches shield can be seen dipping south-southeast and forming the cliff at the beach (Fig. 4.4). The flows are capped with a continuous thick calcrete crust and can be seen only through erosional windows in the calcrete. The red Montaña Bermeja ("vermilion"), a cinder cone of the Recent fissure eruptions series,

Figure 4.4 **(a)** Panoramic view of the Los Ajaches Massif from the top of Montaña Tinasoria. The valley and village of Femés can be seen in the distant background to the left of the photo. On the right-hand side in the distance are the islands of Lobos and Fuerteventura. **(b)** Sequence of basaltic lavas of the Los Ajaches shield volcano, near the town of Yaiza. The lavas in the foreground are from the 1730–36 eruption.

appears attached to the flank of the shield. The small island between Lanzarote and Fuerteventura in the distant background is Lobos, made up of a large basaltic cone and some hydrovolcanic conelets.

Stop 4.3 Femés Return to LZ-702 and turn left to Femés, just in front of a wide breached cinder cone (Caldera Riscada; see Fig 4.9a). The road climbs into the valley of Femés, a northeast-trending barranco carved in the shield, its drainage obstructed by the Caldera Riscada vent and lavas. The road runs over thick layers of soil deposited on the floor of the barranco.

At the top, park at the Mirador de Femés for a magnificent view of the

65

Los Ajaches shield to the east and south of the viewpoint. The northwest wall of the barranco is capped by the Atalaya de Femés, north of the mirador, a large basaltic cone whose lavas flow to the west to cover the marine abrasion platform of El Rubicón. A short walk to the top of the wall of the valley east of the mirador gives an open view of the southern flank of the shield, with dipping lava flows crossed by dykes, U-shape barrancos, and thick alluvial and scree deposits.

Stop 4.4 Salinas del Janubio   From Femés follow the road descending to the village of Las Breñas and the Salinas del Janubio. The lagoon (used in the past as a saltworks) is a circular bay bounded to the east and south by the cliff of the old shield lavas and to the north by the margin of a tongue of lavas emplaced in the 1730–36 eruption. A sand and pebble bank closes the bay to the west.

Walk around the rim of the bay to observe the old basaltic lavas crossed by dykes and the weathered trachytes of the Los Ajaches shield volcano.

Stop 4.5 Punta Papagayo   Continue towards Playa Blanca on the marine abrasion platform covered by Recent-series lavas produced in fissure eruptions; the route is initially on those from the Atalaya de Femés vent and later on those from Montaña Roja. From Playa Blanca, follow the coast to Punta Papagayo, at the foot of the southern flank of the Los Ajaches shield. The track runs through reddish and very altered trachytic lavas crossed by basaltic dykes, similar to those of Salinas del Janubio (stop 4.4). They are probably domes and lavas from differentiated magmas intruding or interbedded in the old basaltic lavas of the Los Ajaches shield.

At the camping ground in Punta Papagayo there is a small bay and beach in which the basaltic lavas and trachytes outcrop.

The Famara Massif

The Famara cliff

As already mentioned, most of the northern and central parts of the Famara shield volcano have been removed by one or more giant landslides and a long period of coastal erosion. The remaining edifice is bounded to the northwest by a 23 km-long and 670 m-high precipitous cliff, the Famara cliff. The cliff, which shows the structure of the old shield (Fig. 4.5a,b) where it is not draped by young lavas (Fig. 4.5c), can be observed from different vantage points (miradores) on the roads that run along the rim (stops 4.6–4.12), as well as the small islands grouped on the shallow marine platform that extends north and west of the shield.

Figure 4.5 **(a)** The Famara cliff with La Graciosa island in the background from the Portillo de Guinate. The lava flows in the foreground are from Los Helechos volcano (stop 4.10). **(b, overleaf)** Sequence of near-horizontal lava flows and old scree deposits of the Famara shield volcano near Punta Fariones. **(c, overleaf)** Lava flows of the Corona volcano cascading down the Famara cliff at Las Rositas. A trail follows the lavas to the coast (stop 4.11).

Stop 4.6 Valle de Femés From Arrecife take the main road to Teguise, the old capital of the island. The road crosses a lava flow from the last phase of the 1730–36 eruption (see Fig. 4.17a). These very fluid tholeiitic pahoehoe lavas extend for more than 20 km to Puerto Mármoles, north of Arrecife, where they flowed for 300 m into the sea.

Past Teguise and the Montaña Guanapay cinder cone topped with an old castle, turn left at km-12 into a dirt track signposted Las Nieves and north into the valley of Femés. The bottom of the valley is filled with lapilli weathered to soil and sediments. At the end of the track, park at the rim of the cliff for a magnificent view of the Famara cliff, with the central part of the island to the west and the cliff and islets to the north.

Stop 4.7 Ermita de La Virgen   About 200 m before reaching the main road again, turn left onto the wide track up hill to Ermita de Las Nieves ("chapel of Our Lady of the snows"). The track follows the rim of the cliff and has many vantage points for views and photographs (best in the morning for the volcanic alignments of the central Recent fissure eruptions and in the

Figure 4.6 **(a)** Panoramic view of the northern islets from the Famara cliff (stop 4.12): La Graciosa, Montaña Clara and Alegranza. **(b)** Oblique aerial view from the south of Montaña Clara and Alegranza.

afternoon for the cliff and islets). Explore the tracks running towards the clifftops for spectacular views.

On arriving at the Ermita de Las Nieves look for a vantage point at the cliff edge to the west, where the entire central part of Lanzarote can be seen on clear days. Towards the north, a magnificent view can be had of the cliff and the three aligned islets (La Graciosa, Alegranza and Montaña Clara), all with hydrovolcanic vents that can be identified by their characteristic yellow tones (Fig. 4.6).

The relatively thin sequence of recent volcanism overlying a flat marine abrasion platform carved on the old Los Ajaches and Famara shield volcanoes is evident from this point. As mentioned before, this is a feature typical of post-erosional islands. The low elevation of the terrain and the presence of a porous and permeable sediment sequence promoted interaction of marine water with most of the eruptions, producing hydrovolcanic explosions and so favouring the presence of large, wide volcanic cones (see p. 75).

Stop 4.8 Mirador at Los Helechos restaurant   Past the Peñas del Chache, the highest elevation of Lanzarote (668 m) topped with a radar dome, the track meets the main road. Park at the restaurant and Mirador de Los Helechos. The Corona volcano can be seen in the distance to the north, as well as the Roque del Este in the sea to the northeast. The flanks of the Famara shield lie to the east and south, with the basaltic flows dipping southeast and the mature U-shape valleys covered with thick scree and alluvial deposits running down to the sea.

Stop 4.9 Mirador Valle del Rincón   In the town of Haría, take the road from the town hall towards the west, following the Valle del Rincón up hill for approximately 1.6 km to the cliff edge, where spectacular views of the bay and central cliff of Famara can be seen from the stone-walled mirador.

Stop 4.10 Mirador de Guinate   Past Haría follow the road north, leaving the Los Helechos volcanic group on the left, and the Corona volcano farther to the right. On arriving close to the Corona volcano, take the road on the left signposted Mirador de Guinate. There is a mirador with a car-park at the end of the valley (less than 1.5 km).

At the top of the 350 m-high cliff, very thick basaltic flows form the interfluves of the present valleys. These flows filled old barrancos and were left as relict hills by **inversion of relief**. The present valleys have their headwalls truncated by the erosive retrogression of the cliff. Lavas of the Los Helechos Group filled the Valley of Guinate and overspilled just south of the mirador cascading down the cliff (see Fig. 4.5c). The lavas, covered with green lichens, have been dated by Hervé Guillou as 72 000 years old. A path descends to the coast over the cascading lava and screes.

To the north there is a closer view of the islets, with the hydrovolcanic cone of Montaña Amarilla on the western edge of the island of La Graciosa immediately opposite.

Stop 4.11 Mirador de Las Rositas   Back on the main road and past the Corona volcano there is a narrow road to the left at the entrance to the village of Yé, which follows the rim of the cliff. A paved path to the left leads to the Mirador de Las Rositas. The valley has been filled with lavas of the Corona volcano. At the parking lot, the pressure ridges at the lobate edge of the flow can be observed (Fig. 4.5c). Walk to the wall of the mirador on the cliff edge to see the lavas falling down to the coast, which can be reached by a well kept path similar to the one at the Valley of Guinate (stop 4.10).

Stop 4.12 Mirador del Rio   The Mirador del Rio is the northernmost mira-dor on the Famara cliff, and toilet and cafeteria facilities are available. The closest view of the islets, especially La Graciosa, can be had from here.

Stop 4.13 Cantil de Fariones   Starting at the port in the fishing village of Órzola there is an excellent view of the east Famara cliff (450 m high) as far as Punta Fariones, made of near-horizontal basaltic lavas with interbed-ded pyroclastic cones and several local discordances. Scree deposits cover the base of the cliff, and dykes cut the lavas. A track from Órzola reaches the foot of the cliff. Park and walk along the trail that runs over the screes at the footwall, but take care because it can be slippery and unstable in places and there is also a danger of rockfalls from the cliff.

At the end of the scree a layer of dunes and sediments with fossils appears interbedded in the lava flows of the old Famara shield, dated between 10 and 8 million years old. The fossils include large eggshells: it was suggested that these were from large non-flying birds of the ostrich type and thus were evidence for "continental bridges" connecting the Canaries to the African continent. However, later studies proved the eggs to belong to pelicans or similar large flying birds: as discussed in Chapter 1, the Canary Islands seem to be entirely oceanic in character.

The Corona volcano and lava tube

From Órzola (stop 4.13) follow the road up hill directly southwards along the foot of the Famara shield and the edge of the Corona lava flows. The road climbs over La Quemada volcano lavas, which are partially covered with sediments and scree deposits, whereas the more recent lavas of the Corona volcano overlie the sediments. On the left-hand side of the road, very large accretionary blocks of lava (Peñas de Tao) drift on the surface of the Corona lavas.

Stop 4.14 Base of the Corona and beginning of the lava tube   Past the La Que-mada cone and lavas (clearly underlying the Corona lavas), park just before the crossroads to observe the lava flow of the Corona volcano that

71

produced the 7.5 km-long and up to 35 m-high lava tube (one of the world's largest), and the natural "skylights" (locally called jameos) produced by collapses in the roof of the tube (Fig. 4.7). Only experienced cavers should venture inside. The end section of the tube (Cueva de Los Verdes and Jameos del Agua) is open to the public (on payment of entry fee), and the usual visitors' facilities are available.

Figure 4.7 **(a)** Lava tube of Corona volcano (in the background), from the Jameo ("skylight") de La Puerta Falsa. **(b)** Interior of the Corona volcano lava tube.

Stop 4.15 Cueva de Los Verdes and Jameos del Agua Descending to the Corona lava field, take the narrow road to the left to the Cueva de Los Verdes ("cave of the shades of green", because of the lichens covering the walls of the volcanic tube). The road goes past the large Jameo de la Puerta Falsa on the way to the entrance. Down at the coast, the Jameos del Agua allows access to the beginning of the submerged part of the tube. A restaurant, swimming pool and a large auditorium have been constructed inside.

Relatively large lava tubes are known to be formed on the sea floor at depths of several thousand metres as a result of submarine eruptions. However, when basaltic lavas flow from land into water, they build lava deltas out from the shore, the lower part consisting of hyaloclastite breccias and pillow lavas, capped with flat-lying massive lavas. The transition marks the sea level at that time (see stop 5.34).

However, as already mentioned, the Corona tube extends for at least 1.6 km under the sea, to a depth of 80 m below present sea level. How could the Corona tube have formed in such a situation? The answer, suggested by the new age data on the volcano, may simply be that the tube formed in subaerial conditions and was later submerged (Fig. 4.8). During the most recent glaciation, from about 70 000 to 15 000 years ago, the sea level was 70–80 m lower than at present. The lavas of the Corona volcano were formed about 53 000 years ago and so would have flowed on a 2 km-wide coastal abrasion platform covered by Los Helechos lava flows, until they reached the shoreline (Fig. 4.8: stage 1). The rising sea level after the most recent glaciation submerged the final part of the tube to its present conditions (Fig. 4.8: stage 2).

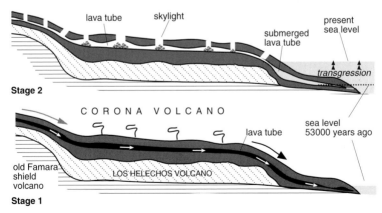

Figure 4.8 The formation of the Corona volcano lava tube, later submerged as sea level rose at the end of the last glaciation. Stage 1: original situation; stage 2: present day.

The northern islets

Follow the road from the Jameos del Agua along the coast on lavas of the Corona volcano. Near Órzola, several patches of calcareous aeolian sands contrast strongly with the basaltic lavas.

The short boat trip to La Graciosa provides an opportunity for spectacular views of the Famara cliff, with massive screes on both sides of Punta Fariones (see Fig. 4.4a).

Stop 4.16 The island of La Graciosa   On the shallow marine platform north of Lanzarote lie several small islets and rocks, namely La Graciosa (27 km^2), Montaña Clara (1 km^2), Alegranza (12 km^2) and Roques del Este and del Oeste ("east and west rocks"). Only La Graciosa is inhabited. All were formed by relatively recent basaltic eruptions of similar age to the recent fissure eruptions at the centre of the island.

The island of La Graciosa (see Figs 4.5a, 4.6a) can be explored in a day (bicycles are available), but it may be worthwhile to stay longer to relax and swim in the beautiful transparent waters of the shallow strait of El Rio, which separates La Graciosa from the main island. The harbour and village of Caleta del Sebo afford spectacular views of the cliffs of Famara, with their impressive piedmonts undergoing erosion (see Fig. 4.5a).

A track crosses the island between the cinder cones (Montaña del Mojón and Montaña Pedro Barba) to allow a close view of the islets of Montaña Clara and Alegranza. Montaña Amarilla, at the southernmost tip of the island, shows interesting features of hydrovolcanic eruptions. From the top of this cone the entire Famara cliff can be observed, with the cascades of lavas from the Los Helechos and Corona volcanoes and the thick, extensive screes at the foot of the cliff.

Recent volcanism

The central fissure eruptions

The central saddle of Lanzarote, between the higher shield volcanoes, is crossed by northeast–southwest-trending chains of volcanic cones, from which issued a multitude of lava flows. These chains of vents originated in different fissure eruptions from the Pleistocene onwards, most recently in 1730–36 and 1824. Most of the cinder cones are breached, with the openings in the direction of the dominant winds during the eruption, generally the northeast trade winds.

Stop 4.17 The Montaña Guardilama viewpoint One of the best vantage points to observe the central volcanic lineations is the Montaña Guardilama, the highest cinder cone in Lanzarote (603 m). From Arrecife (or the airport) follow the main road west to Yaiza. Pass the quarries described in stop 4.1 and turn right just after km-12 (at Los Llanitos) into a narrow road to La Asomada, at the foot of the volcanic lineation. As the road starts climbing, turn right and follow the foot of the volcanoes until arriving at the saddle between montañas Guardilama and Tinasoria.

Park and walk to the top of Montaña Guardilama for a spectacular southwesterly view (best in the morning) of the Los Ajaches shield volcano (west and southwest), with the islands of Lobos and Fuerteventura in the distance (Fig. 4.4a) and the large cinder cone of Caldera Riscada in the foreground (Fig. 4.9a). To the west and northwest there is an open panoramic view of the 1730–36 vents and lava fields. The crater of Montaña Tinasoria is occupied by vineyards grown in hollows in the lapilli (Fig. 4.9b), a technique developed after the 1730–36 eruption and described below. The short walk to the summit of Montaña Guardilama (183 m from the saddle) takes 20–30 minutes. The effort is rewarded with the view of the entire central plain of the island. In the mornings on clear days the entire area covered by the 1730–36 and 1824 eruptions can be observed, as well as the vents of preceding fissure eruptions, lighter in colour and frequently encircled by the historical lavas. In the afternoons the view extends to the northeast to the distant Famara shield and cliff, and to the

Figure 4.9 Different volcanic cones of the post-erosional fissure volcanism: **(a)** Caldera Riscada, near the town of Yaiza. **(b, overleaf)** Montaña Guardilama, the highest cinder cone in Lanzarote (603 m), one of the best vantage points from which to observe the central volcanic lineations of Lanzarote. In the foreground are vines cultivated in lapilli, with the characteristic walls to protect the plants from the dominant trade winds (stop 4.17). **(c, overleaf)** Volcanic cones in the central plain of the island.

northern islets. Directly east lies the Atalaya de Femés–Guardilama–Tahiche volcano chain, the highest and largest cones produced by the Pleistocene fissure eruptions of the island.

The hydrovolcanic eruptions

In the volcanic landscape of Lanzarote the size and shape of the volcanic cones are particularly interesting, being generally larger and flatter and

with wider craters than in the remainder of the Canaries. This morphology is typical of eruptions in which marine water interacts with magma to produce hydrovolcanic eruptions. The low elevation of the island after erosion, often below 300 m, has favoured hydrovolcanism, generating frequent tuff cones and tuff rings. Many of the eruptive vents can be observed to have started as hydrovolcanic and end as Strombolian, leading to the large wide-crater volcanic cones. This explains the frequent use of the toponym "caldera" in the recent fissure eruptions of Lanzarote, although the features thus named are actuallynot calderas in the modern sense but volcanic cones.

Although **tuff cone**, tuff ring and **maar** are terms commonly used as synonyms, maar (from the Eifel area in Germany) should be used only for tuff cones and rings with craters excavated deeper than the pre-existing ground surface and so frequently filled with water (e.g. Montaña del Golfo, stop 4.18). The difference between tuff cones and rings is morphological: tuff rings have larger crater diameters and lower rim heights.

Stop 4.18 The maar of El Golfo   From the town of Yaiza follow the road to the Salinas del Janubio (stop 4.4), crossing the lava flows of different phases of the 1730–36 eruption and running around the sub-horizontal sequence of lavas of the old shield of Los Ajaches (Fig. 4.4b). On arriving at the junction, turn right to the road towards El Golfo, following the coast over lava flows from the 1730–36 eruption (note the accretionary lava balls typical of lavas that have flowed over long distances).

Past Montaña Bermeja (a red Strombolian cone with an initial hydrovolcanic phase) follow the road to a parking lot at the foot of Montaña del Golfo. A short walk leads to the crater, excavated by hydrovolcanic explosions to a depth below sea level. As a result, it is filled with highly concentrated salt water colonized by green algae (Fig. 4.10a). In the walls of the cone interesting features of hydrovolcanic eruptions (palagonitic tuffs, cross bedding, bomb sags, abundant shattered olivine xenoliths, etc.) can

Figure 4.10 **(a)** Aerial view from the south of Montaña de El Golfo, encircled by lava flows from the 1730–36 eruption (stop 4.18). **(b)** View inside the explosion crater of the maar of Montaña de El Golfo (stop 4.18).

be closely observed (Fig. 4.10b). The top of the cone (the late eruptive phases) are Strombolian, evidence of the isolation of the crater from the sea as the cone developed.

On leaving the parking lot, turn right at the first and second junctions to the fishing village of El Golfo. Here is a different view of the crater of Montaña del Golfo.

Return to Yaiza using a direct road crossing a Pleistocene chain of cinder cones (Montaña Gabriela–Pico Redondo) completely encircled by the 1730–36 lavas.

Stop 4.19 The tuff cone of Caldera Blanca   The 1125 m-diameter crater floor of this spectacular 1750 m-wide 310 m-high tuff cone lies above the pre-existing ground surface and is dry. The cone is colonized by lichens, which give it its conspicuous white colour and hence its name (Fig. 4.11a,b).

From Tinajo take the road to El Islote. Pass the Montaña de Tinajo and Montaña de Tenezar cinder cones, the latter partially made of indurated hydrovolcanic lapilli, which has been quarried to extract blocks for building purposes. After 2.5 km of dirt track, turn left twice to a group of houses (El Islote) with the Caldera Blanca in front. Follow the track to the foot of the mountain and a trail ascending to the top.

The view from the rim of the tuff cone (460 m above sea level) covers most of the 1730–36 eruption volcanic alignment and lava fields (best viewed in the morning) and the volcanic cones crossing the centre of the island (best viewed in the afternoon). Distant in the south, the Los Ajaches smooth profiles of the old shield volcano are visible (best viewed in the mornings). To the far northeast, the Famara shield and the northern islets can be seen and photographed in the afternoons.

The Caldera Blanca can be easily walked around at the top in about an hour, giving spectacular panoramic views of the island. At the eastern side another smaller tuff cone (Montaña Caldereta) emerges, encircled by lavas of the Pico Partido volcanic group (stop 4.21).

Stop 4.20 The El Cuchillo maar   Take the road north from Tinajo to Las Vegas and turn left at km-1 to the village of El Cuchillo ("the knife", because of the sharp edge of the tuff ring). Take a narrow road to the village of Sóo, which circles the rim and crosses the crater of the volcano, one of the best examples of a tuff ring in the Canaries.

Partially covered with lavas, the cone is at least 2750 m wide and 180 m high (Fig. 4.12). The floor of the 2375 m-wide 130 m-deep crater is only 40 m above present sea level, and was formed in at least two different eruptive phases, the most recent one giving place to the deepest western explosion crater.

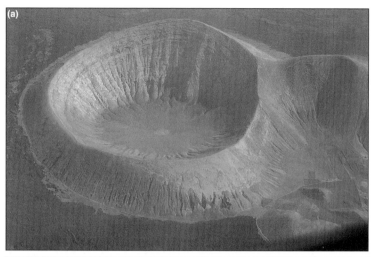

Figure 4.11 **(a)** Aerial view from the northeast of the large tuff cone of Caldera Blanca (stop 4.19). **(b)** Panoramic view of Caldera Blanca from the southern tip of the crater rim (stop 4.19).

The walls of the cone are made of thinly bedded gravelly palagonitic tuff breccia, with abundant peridotitic xenoliths and angular blocks of the pre-existing rock (foreground in Fig. 4.12). Features typical of high-energy hydrovolcanic explosions are abundant: cross bedding, ballistic impact sags, and so on. The surface is very slippery and it is therefore safer to observe details of these features from the floor of the crater.

Figure 4.12 Panoramic view of the southern wall of the 2750 m-wide and 180 m-high crater of El Cuchillo ("the knife"), one of the best examples of a tuff ring in the Canaries. The layers correspond to the hydrovolcanic explosions. In the foreground, angular blocks of the host rock shattered by the explosions can be seen (stop 4.20).

The historical eruptions

The 1730–36 volcanic eruption

The basaltic eruption that took place in Lanzarote between 1 September 1730 and 16 April 1736 produced about 30 volcanic cones aligned along a 14 km volcanic fissure. The eruption covered about 25 per cent of the island with lapilli and lava flows.

This eruption was exceptional in terms of the "normal" style of the historical volcanism of the Canaries, because of its duration (2056 days, against an average in the Canaries of 38 days) and extent (about 215 km², average 3 km²). The effects were devastating inasmuch as the major part of the most fertile land was destroyed, along with 26 villages, and the subsequent famine eventually forced the majority of the population to leave the island.

The 1730–36 eruption, one of the largest historical basaltic fissure eruptions, has attracted scientific interest since the activity ceased (Fig. 4.13a). Many descriptions and detailed accounts of the eruption have been published, but two eye-witness accounts are especially noteworthy: the

81

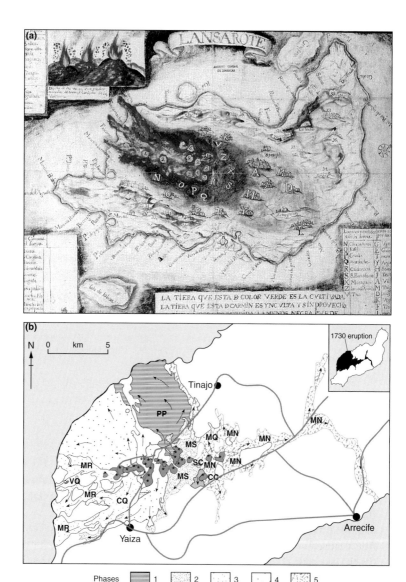

Figure 4.13 **(a)** In November 1730 the governor of Fuerteventura sent an artist to Lanzarote to "precisely map the places lost by lavas, those lost by sands (lapilli) . . .". The original oil painting is in the National Archives in Simancas (from Carracedo & Rodríguez Badiola 1991). **(b)** Simplified geological map of the 1730–36 eruption indicating the main phases of the eruption (from Carracedo et al. 1992).

CC: Caldera de Los Cuervos PP: Pico Partido SC: Santa Catalina MS: Montañas del Señalo VQ: Volcán El Quemado MR: Montaña Rajada CQ: Calderas Quemadas MF: Montañas del Fuego MN: Montaña de Las Nueces MC: Montaña Colorada

detailed entries in the diary kept by the parish priest of Yaiza (included in the work of von Buch 1825) and the official reports presented to the Royal Court of Justice of the Canary Islands by the local authorities appointed specifically to manage the crisis, the original documents being found on file in the Spanish National Archives in Simancas.

This eruption greatly reduced economic activity, and most of the population left the island. However, upon their return, a newly developed agricultural practice based on mulching with lapilli (to trap atmospheric moisture and so irrigate the crops *in situ*) made the economy flourish and the population doubled in a few years after the eruption.

An interesting and highly unusual feature of the 1730–36 eruption was the emission of lavas ranging in composition from basanites and alkali basalts to **tholeiite** basalts. The latter are a predominant rock type in the shield stage of growth of many oceanic islands, notably the Hawaiian Islands. However, they are very rare in the Canaries, and their occurrence in the 1730–36 eruptions suggests that the source of the lavas lay in an unusually heterogeneous region of the mantle.

1.The initial eruptive episodes
Geological studies and analyses of the eye-witness accounts allowed the definition of five different stages in the 1730–36 eruption (Table 4.2, Fig. 4.13b).

From 1 September 1730 to the beginning of 1731, three main eruptive centres opened in an alignment along a northwest–southeast fissure. Had the eruption ended at the end of this stage, it would have been similar to the normal run of eruptions in the Canaries.

The initial vent of 1730 has been generally located and associated with the area of thermal anomalies at the Islote de Hilario (stop 4.24), except by Leopold von Buch (1825), who indicated an area close to the Caldera de Los Cuervos, the most probable initial vent. The account of the parish priest of Yaiza listed the villages destroyed in the first days of the eruption and described the sudden change in direction of the lava flows caused by an "obstacle". The above-mentioned cinder cone best fits the account. The obstacle may correspond to a portion of the wall of the crater that was dragged by the lavas 250 m northwards, this being large enough to divert the flow (Fig. 4.14a). The Caldera de Los Cuervos lavas, very fluid, poor in silica and abundant in olivine inclusions, were emitted from 1 to 19 September 1730, flowing north and west to the coast to cover a large area and razing several villages.

On 10 October 1730 two new eruptive vents opened: Pico Partido (stop 4.23) and Caldera de Santa Catalina (Fig. 4.14b). Until February 1731, large amounts of lapilli and lava flows covered an extensive area

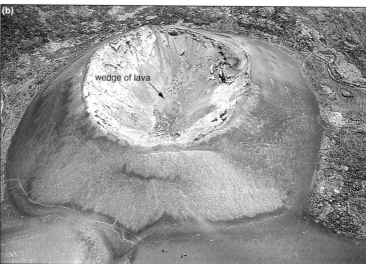

Figure 4.14 **(a)** Caldera de Los Cuervos (from Montaña Negra), the initial vent of the 1730–36 eruption. The account of the parish priest of Yaiza quotes a sudden change in the direction of the lava flows from this vent caused by an "obstacle", most probably the large portion of the crater rim (arrow) detached and towed by the lavas (stop 4.21). **(b)** Caldera de Santa Catalina, a phase 1 vent that erupted large volumes of pyroclasts that destroyed many villages and farms (stop 4.21). **(c, opposite)** Volcanic vents of the initial phases (1 and 2) of the eruption (Pico Partido and Montañas del Señalo), from Montaña Negra. The itinerary proposed in stop 21 is shown.

to the south, destroying more farmland and villages and threatening the main village of Yaiza. These lavas were extremely rich in olivine inclusions, which occasionally form 30–40 per cent of the rock.

2. The Montañas del Señalo volcanic group

At this stage the main N80°E fissure may have opened, producing a series of vents with a constant west-to-east progression between March and July 1731 (Fig. 4.14c). The lavas were more silica saturated in composition (olivine basalt evolving to olivine tholeiites) and with few olivine inclusions; the higher magma viscosities and lower average effusion rates in this phase resulted in shorter flows that stopped half way to the coast.

Once the main fissure was established, the eruption continued for another five years. This feature may be ultimately responsible for the anomalous character of this eruption.

3. Volcán del Quemado–Montaña Rajada–Calderas Quemadas volcanoes

An abrupt change in the behaviour of the eruption is indicated by the apparently sudden jump in the focus of eruptive activity of at least 12 km to the westernmost edge of the fracture. At the end of June 1731, a submarine eruption seems to have started on the west coast, where the trend of the vents enters the sea. Hydrovolcanic explosions at the coast, and probably even deeper submarine eruptions that brought to the shore species of apparently deepwater fish previously unknown in the area, are reported in the diary of the priest of Yaiza. The first eruptive centre on land was El Quemado volcano (stop 4.18), a cluster of small elongated cinder cones and hornitos emplaced in the fracture and located 1 km from the coast (Fig. 4.15a). The eruptive activity migrated continuously to the east along the fracture, forming a close alignment of cinder cones: Montaña Rajada, with a spectacular lava channel (Fig. 4.15b) and the four Calderas Quemadas (Fig. 4.15c). Lavas of a composition and evolution similar to the preceding phase (alkali basalts to olivine tholeiites) continued to flow seawards, spreading over a large area of the western part of the island.

Lava flows from Montañas Quemadas destroyed the farmland of Yaiza, almost reaching the town, and the priest left the island, together with the other inhabitants of the village (von Buch 1825).

4.Montañas del Fuego volcanic group

Early in 1732, the eruption seems to have stopped its previous continuous progression along the general N80°E trend and became stationary for a long period. Volcanic activity was concentrated in a small area around the Timanfaya volcano, forming the complex volcanic edifice of Montañas del Fuego by overlapping cinder cones (stop 4.24). The eruptive style changed and lavas were emitted from clusters of hornitos and large lava tubes and channels (Fig. 4.16a). These lavas varied from alkali basalts to highly evolved olivine tholeiites.

5.The final episodes

The next record is an account of the onset of this fifth and last phase of the eruption in March 1736.

In the short time that elapsed between mid-March and 16 April 1736, the eruption abruptly ended with two new emission centres: the Montaña de Las Nueces ("mountain of the walnuts", because of the crackling noise caused by walking on the lavas) and Montaña Colorada volcanoes. Very fluid tholeiitic lavas issued from the Montaña de Las Nueces volcano, probably at very constant rates, entering the sea near Arrecife (at Puerto Mármoles), at a distance of more than 20 km (Fig. 4.17a).

This last episode probably lasted only 10–15 days, as volcanic activity in the island is known to have ceased completely on 16 April 1736 (Fig. 4.17b).

To observe in detail and photograph the main features of the different episodes of the 1730–36 eruption it is preferable to keep outside the limits of the national park, for the reasons already explained. Accordingly, the stops devoted to the analysis of the eruption are outside the park, with the exception of stop 4.24, which comprises the visit to the thermal anomaly area of Islote de Hilario and the official guided circuit.

Figure 4.15 **(a)** Aerial view from the west of the pre-1730 eruption volcanic lineation (Montaña de Juan Perdomo to Montaña Encantada), with the 1731 eruption vent of El Quemado volcano (centre of the photograph) encircled by lava flows of phase 3 of the eruption. **(b)** Aerial view from the west of the large lava channel of Montaña Rajada, a vent of the Montañas del Fuego Group, phase 4 of the eruption (stop 4.24). **(c, overleaf)** Aerial view from the northwest of the Calderas Quemadas alignment, from phase 3 of the eruption (stop 4.24).

Stop 4.21 The initial vents   From Arrecife, take the road to San Bartolomé and at the monument to El Campesino crossroads turn west onto GC-730. Past Masdache turn north at km-15 and park at km-1 at the entrance to a track leading to Caldera de Los Cuervos volcano, 500 m to the west.

Walk around the volcano, which overlies an older cinder cone with the lapilli partially transformed into yellow soil. Go into the crater to see the lava conduit, with abundant olivine xenoliths.

Return to the road and at km-2 turn west onto another track, which follows the entire initial volcanic lineation of 1730–36 (about a three-hour walk). Make arrangements to have a car waiting at the entrance to the national park where the trail joins the main road.

The trail passes the Caldera de Santa Catalina, an explosive eruptive vent, possibly the one responsible for most of the thick pyroclastic layer that destroyed farms at the beginning of the eruption (Table 4.2). A wedge of lava intrudes at the bottom of the crater (see Fig. 4.14b).

The trail encircles the south flank of the volcano and crosses a cluster of hornitos at the southern flank of the Montañas del Señalo volcano. The trail crosses collapse structures in the flank of the volcano and a spectacular pahoehoe lava field.

The trail turns northwards into a saddle between the montañas Rodeos and Miraderos (west) and Montañas del Señalo (east). From this point there is an impressive view (best in the morning) of the Caldera del

Figure 4.16 **(a)** Aerial view from the northwest of the Islote de Hilario and the hornitos and lavas of the Montañas del Fuego Group (stop 4.24). **(b)** The Caldera del Corazoncillo, a vent of the early stages of the 1730–36 eruption (oblique aerial view from the south).

Figure 4.17 **(a)** Lava flow of the Montaña de Las Nueces vent (phase 5), one of the late eruptive centres of the eruption (stop 4.22). This very fluid tholeiitic lava flow entered the sea near Arrecife (at Puerto Mármoles), at a distance of more than 20 km. **(b)** Montaña Colorada cinder cone, the last eruptive vent of the 1730–36 eruption (stop 4.22), seen from the top of Montaña Negra.

Corazoncillo, covered with oxidized (red) lapilli (Fig. 4.16b), possibly an eruptive vent of the initial stages of the eruption (coeval with Pico Partido?). Farther east lies the volcanic lineation of the intermediate stages of the 1730–36 eruption (Montañas del Fuego).[1]

The trail drops to the main road near the entrance to the national park, crossing lavas of the initial stages of the eruption and leaving the Pico Partido volcanic group on the eastern side (see Table 4.2 and stop 4.23).

Stop 4.22 Final vents: Montaña de Las Nueces and Montaña Colorada Park at the entrance to the track at the beginning of the long walk described in stop 4.21 and walk to the nearby Montaña de Las Nueces volcano. This vent and lavas of the 1730–36 eruption (see Table 4.2) have been almost completely destroyed to extract building materials. This is a consequence of the absurd delimitation of the national park, which excluded a relatively small but most important area of the 1730 eruption, later included in the Parque Natural de Los Volcanes. However, local protection laws are far less stringent and the Parque Natural is not as effective in terms of protection as the Parque Nacional is (this applies to the other Canaries as well).

Lavas issuing from this vent covered large areas, forming spectacular lava fields with lava lakes and streams, lava tubes (Cueva de Los Naturalistas), pressure ridges and tumuli. They flowed for long distances towards the villages of Masdache and Mozaga, entering the sea north of Arrecife (see Figs 4.13b, 4.17a). On the eastern side of the road stands Montaña Colorada, the last vent of the 1730–36 eruption (Fig. 4.17b), whose lavas flowed towards the north coast (see Table 4.2, Fig. 4.13b).

Stop 4.23 The Pico Partido group The Pico Partido group (see Table 4.2, Fig. 4.13b) is mentioned in stop 4.21. The perfectly preserved and splendid volcanic features (with lavas cascading and filling craters to form en-echelon lava lakes, lava channels, hornitos and pahoehoe lava fields) deserve attention. At km-9 on the road from Yaiza to Tinajo through Montañas del Fuego, park at the "devil marker" close to a cinder cone (Montaña Tíngafa) and look for a barely visible, very narrow trail in the lavas on the opposite side of the road, leading directly towards Pico Partido (Fig. 4.18). Past the lavas from Montañas del Señalo the trail enters the lapilli field of the Pico Partido group. Avoid walking over the thin pahoehoe lavas, because they fracture easily.

Walk to the lava channel and inside the lava lakes (best for photography in the afternoon). Olivine xenoliths, some the size of a football, are still

---

1. The book *Lanzarote: la erupción volcánica de 1730*, edited in 1991 by the Cabildo de Lanzarote, includes a colour geological map of the eruption, on the back of which is a map of the main trails and stops for detailed observation of the 1730–36 eruption.

Figure 4.18 Aerial view from the north of the Pico Partido group (phase 1), with spectacular lava lakes and lava channels. The white surface is formed by lichens colonizing the lavas (stops 4.21 and 4.23).

extraordinarily abundant, despite the numbers already sold as holiday souvenirs.

Stop 4.24 The Montañas del Fuego   The Montañas del Fuego Group, corresponding to the prolonged quasi-stationary intermediate stage of the eruption (Table 4.2), lies almost entirely inside the national park and can be seen only from coaches on the official tour circuit.

The main parking lot and restaurant are at the Islote de Hilario, a pre-1730 cinder cone (Fig. 4.19a). Temperatures up to 600°C at 10 m depth are high enough to burn wood at outcrop. This dry thermal anomaly is not exploitable except to maintain artificial geysers (Fig. 4.19b) and a naturally heated grill to impress tourists.

The occurrence of these high temperatures is related to a slowly cooling buried mass of lava remaining from the 1730–36 eruption and the emission of very hot magmatic gases through fissures (Fig. 4.19c).

Spectacular volcanic features can be observed and photographed during the coach trip around the park (see Fig. 4.16).

ROGER JONES

Figure 4.19 **(a)** The Islote de Hilario, a pre-1730 cinder cone encircled by vents and lava flows of the 1730–36 eruption. **(b)** The dry thermal anomaly (600°C at 10 m and > 400°C at outcrop) of the Islote de Hilario sustains a restaurant and artificial geysers among the island's main tourist attractions. **(c, overleaf)** Sketch explaining the origin of the surficial thermal anomaly, related to a deep core of slowly cooling lava from the 1730–36 eruption. In the absence of water, the heat is transmitted by atmospheric air mixed with small volumes of very hot volcanic gases (stop 4.24).

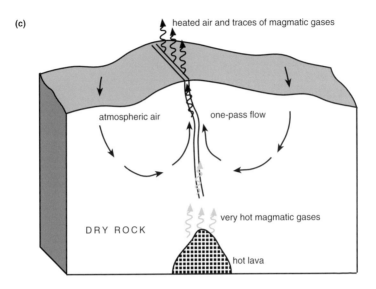

The 1824 eruption

Almost a century later, in July 1824, another eruption took place, this time with the normal features of eruptions in the Canaries. There were three well separated emission centres, aligned along a fissure (possibly a dyke) running east-northeast–west-southwest for more than 13 km.

The initial vent, Tao volcano, opened on 31 July 1824, at the east end of the alignment. This vent has been completely destroyed. The last one, Tinguatón, opened on 16 October, at the centre of the fracture. Between these two in time, on 29 September, the spectacular Volcán Nuevo del Fuego ("new volcano of fire") formed at the west end of the alignment, this being the only one inside the Timanfaya National Park today. Ironically, this volcano, with some of the most spectacular aa and pahoehoe lava fields and lava channels in the entire archipelago, was quarried to extract materials to make the road inside the park and is not included in the circuit.

Stop 4.25 Tinguatón volcano   From stop 4.24, pass km-10 and turn right into a short track to the foot of Tinguatón. Climb inside the crater to view several open circular drainage sinks (proceed with extreme caution: the openings are very deep). The accounts state that these vents released jets of hot water that breached the cone, eroding a narrow gap in its northeastern flank: this is a further manifestation of the abundance of groundwater near the surface. This volcano, the last eruptive activity in Lanzarote, produced a short tongue of lavas that blends in with the visually very similar 1730–36 lavas.

Chapter 5

Gran Canaria

Introduction

Like the other older Canary Islands, Gran Canaria (Fig. 5.1) shows two main periods of igneous activity, of Miocene (14–15 to about 8 million years ago) and Pliocene to Holocene age (5.5 million years ago to sub-historic), corresponding to the shield-stage and post-erosional periods characteristic of other ocean-island volcanoes. The sequence of volcanic activity and periods of quiescence are summarized in Table 5.1 Figure 5.2. The two phases have combined to produce some of the most spectacular and complex scenery in the Canaries, together with intense erosion before and during the continuing post-erosional volcanism, which although impressive has never succeeded in drowning the well developed drainage systems for long. Another consequence of this long history of volcanism

Table 5.1 Simplified history of volcanic activity on Gran Canaria, showing division into Miocene shield-building volcanism, ending with the formation, filling and resurgence of the caldera; and Pliocene to recent post-erosional volcanism, divided into several episodes of activity.

Period of activity	Radiometric ages of main periods of activity	Major stratigraphic units (ages where defined)	Important subdivisions	Comments
Post-shield volcanism (Pliocene to Recent)	5.5 million years - 5000 years (volcanism must be assumed to be continuing into the future)	Plio-Pleistocene to recent volcanic field (<3 million years) Roque Nublo Group (5.5 to 3 million years)	Llano de la Pez formation	Recent volcanic field in NE of island Nephelinite lavas in centre of island Stratovolcano with multiple lateral collapses and low-temperature ignimbrites
Erosional hiatus	No preserved volcanic or intrusive rocks: various valley fill and alluvial fan sedimentary units, plus coastal marine sediments			Period of intense erosion, producing deep barrancos
Shield-stage volcanism (Miocene)	More than 14.5 to 9.0 million years	Felsic Group (14 to 9.0 million years) Basaltic Group (14.5 to 14 million years)	Fataga formation (+Intracaldera intrusive complex) Mogán Formation Hogarzales Formation Guigui formation	Caldera formation c. 13 million years, accompanied by ignimbrite eruptions Basaltic shield volcanoes

Figure 5.1 Shaded-relief topographic image of Gran Canaria. Note in particular the radial pattern of the deep barrancos incised into the rocks of the island. (Image: GRAFCAN).

and erosion is that the distribution of volcanic rocks is also complex, with small erosional remnants or **outliers** of younger rocks perched high on the sides of the major canyons, forming mesas within the canyons, or capping the intervening plateaux or cumbres; other even younger rocks partially fill other valleys, especially in the north of the island. However, a broad distribution of the main groups of rocks can be defined as follows (see also Fig. 5.2).

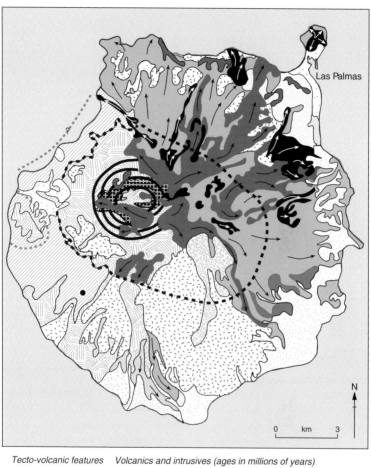

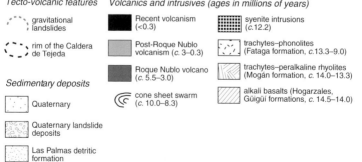

Tecto-volcanic features

- ╱‾‾╲ gravitational landslides

- ╭ ‾ ‾ ╮ rim of the Caldera de Tejeda

Sedimentary deposits

- Quaternary

- Quaternary landslide deposits

- Las Palmas detritic formation

Volcanics and intrusives (ages in millions of years)

- Recent volcanism (<0.3)

- Post-Roque Nublo volcanism (c. 3–0.3)

- Roque Nublo volcano (c. 5.5–3.0)

- cone sheet swarm (c. 10.0–8.3)

- syenite intrusions (c.12.2)

- trachytes–phonolites (Fataga formation, c.13.3–9.0)

- trachytes–peralkaline rhyolites (Mogán formation, c. 14.0–13.3)

- alkali basalts (Hogarzales, Güigüí formations, c. 14.5–14.0)

Figure 5.2 Geological map of Gran Canaria. The complexity of the distribution of the rocks partly reflects the deep erosion that affected the island after the end of shield-building activity and has continued to the present (modified after Carracedo et al. 2002).

Shield-building volcanism (Miocene)

Rocks formed during the Miocene dominate the south and west of the island, and can be divided into two parts, respectively dominated by basaltic lavas and alkaline felsic rocks (trachytes, phonolites and rhyolites).

The older basaltic rocks (the Hogarzales and Güigüí formations) form overlapping shield volcanoes in the west of the island and are cut by dyke swarms in several areas. Similar rocks probably form most of the submarine volcanic rocks on which the island rests, but, unlike several of the other islands, Gran Canaria lacks an uplifted seamount sequence. Again, unlike the other islands, there is no clear onshore evidence for the occurrence of large-scale volcano lateral collapses during the stage of shield building, although the steep curved cliffs south of Agaete on the northwest coast are very suggestive of a collapse-scar headwall, and a similar embayment exists off shore.

As on Tenerife more recently, the later stages of shield-stage activity on Gran Canaria involved development of shallow magma bodies in which felsic magmas formed, large-volume explosive volcanic eruptions and caldera collapses. The Mogán and Fataga formations are alkaline felsic rocks (Mogán: trachytes to **peralkaline** rhyolites; Fataga: trachytes to phonolites) associated with the formation and subsequent filling and resurgence of a large caldera in the centre of the island (Fig. 5.2). They formed in a series of large-volume pyroclastic eruptions; most deposits within these formations appear to be lavas, but are actually very strongly welded ignimbrites that flowed like lavas after deposition at very high temperatures (**rheomorphic ignimbrites**). The main caldera collapse occurred during emplacement of the oldest Mogán, formation ignimbrite and the overall sequence is much thicker (over 1 km) inside the caldera than outside (overall thickness around 500 m). Subsequently, the centre of the caldera was uplifted by emplacement of a large intrusive complex (Fig. 5.3) composed of many minor intrusions, especially abundant inclined intrusive sheets forming a cone-sheet swarm centred close to Tejeda. Intense alteration also took place, resulting in the bleached to bright-red coloration characteristic of the rocks in this area. Repeated eruptions from these intrusions produced the younger pyroclastic rocks and lavas of the Fataga formation.

After the end of Miocene volcanic activity, the volcanic sequence was deeply eroded, producing an apron of sedimentary rocks, mainly alluvial fan conglomerates (best seen around Arguineguín and Las Palmas), around a mountainous central highland with a radial pattern of barrancos. This topography has been partly preserved beneath younger Pliocene rocks and is remarkably similar to the present-day topography, with several barrancos having been re-incised close to their original positions.

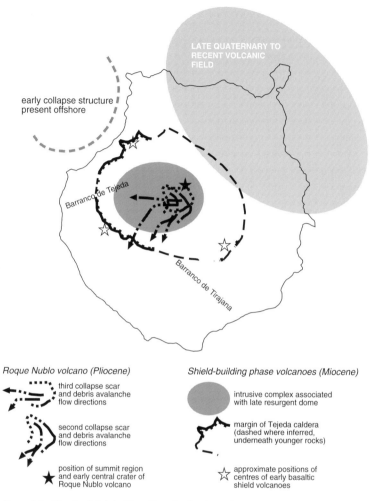

Figure 5.3 Structural sketch of Gran Canaria, showing major craters and volcanic centres and the relative positions of the three principal subdivisions of the volcanic activity: the Miocene shield-building volcanism, the Pliocene Roque Nublo volcano with its lateral collapses, and the Pliocene to recent volcanic field.

Post-erosional (Pliocene to recent) rocks

Post-erosional rocks on Gran Canaria were formerly divided into two discrete volcanic series separated by a period of inactivity and a marked change in composition, but more recent studies indicate that there was continuous activity throughout this period and intense contemporaneous erosion (hence abundant unconformities including deeply eroded palaeo-valleys), and that the entire spectrum of magma compositions was present during the earlier stage of activity. The apparent break is attributable to a decline in the magma supply rate and the disappearance of the shallow magma chamber that fed the Roque Nublo volcano.

From about 5.5 to 2.7 million years ago the mountainous centre of Gran Canaria was occupied by a large stratovolcano, Roque Nublo, on the flanks of which there were many small basaltic vents. The volcano has been intensively studied in recent years, partly because its southern and western flanks were affected by lateral collapses (Fig. 5.3), producing well exposed debris-avalanche deposits. Detailed mapping of the collapse structures has shown that a sequence of at least three lateral collapses occurred, separated by periods of re-growth of the volcano. The collapses were of only moderate size, with volumes of the order 20–70 km^3, but incision of deep barrancos since the Pliocene has produced spectacular exposures of the lower parts of the collapse structures in an area with up to 700 m-high cliffs. These exposures offer many insights into the mechanics of movement and disaggregation of the sliding megablocks that form in the early stages of collapses.

The older rocks of the Roque Nublo volcano are mainly basalt and trachybasalt lavas. Close to Las Palmas, they are intercalated with shallow-water marine sediments (the Las Palmas formation). The onset of growth of the Roque Nublo volcano occurred at about the same time as a global rise in sea level to about 50 m above present in the early Pliocene. The Las Palmas formation sediments now occur up to 110 m above present sea level, indicative of slight uplift in northeastern Gran Canaria, by several tens of metres. This uplift is unusual in the Canary Islands, which are for the most part extremely stable; it may reflect tilting of the island as the ocean crust beneath is uplifted by intrusions that feed the recent volcanic field in this part of the island.

Higher up the Roque Nublo sequence, the rocks are dominated by distinctive ignimbrites with a very high content of lithics. Before erosion, the height of the volcano may have been as much as 3000 m. Various intrusive rocks occur in an area in the centre of the island (Fig. 5.2), which also contains an early volcanic crater-fill sequence. The main collapse deposits and structures occur in the south and west of this area and as valley-filling debris-avalanche deposits farther south and west again. After each collapse, the scar was deeply eroded; following the second collapse the stratovolcano grew again; after the third the stratovolcano was replaced by a much more diffuse alkali basic volcanic field.

Later Pliocene and Pleistocene volcanic rocks cover most of the northeast of Gran Canaria (Figs 5.2, 5.3). They consist of alkali basalt, basanite and **nephelinite** lavas, with lesser amounts of scoria and ash close to the volcanic vents. Perhaps the most notable units are the late Pliocene Llanos de La Pez formation in the centre of the island, which forms the upper part of the great cliff east of Tejeda, and the various hydrovolcanic explosion craters produced by steam explosions where magma came into contact with water-saturated rocks close to the surface, most notably the Caldera de Bandama. In a few places along the northeast coast of the island, these lavas are intercalated with shallow marine sediments, indicating that the slight uplift of this part of Gran Canaria is continuing. The most recent eruption (Montañón Negro) has been dated at about 1500 BC.

Logistics on Gran Canaria

Gran Canaria is densely populated, especially in the north of the island; hence, the bus service is comparatively good. Taxis are expensive, but, as elsewhere in the Canaries, hire cars are plentiful and very cheap. A word of warning: the roads in the mountainous interior are narrow, winding and frequently busy. Inexperienced drivers often find them intimidating, so you may find it preferable to use buses and walk in this area, linking up the walks between locations that are described here.

Almost all accommodation in Gran Canaria is to be found either in the resorts on the south coast between Playa del Inglés and Puerto de Mogán, or in Las Palmas. All these places are noisy and crowded. The southern resorts offer good access to areas in the south of the island, Las Palmas to the north coast. Infinitely preferable to either, as long as you have a car or are prepared to do plenty of walking, are the few small hotels, pensions and apartments in Agaete, San Nicolás de Tolentino, San Bartolomé de Tirajana and Tejeda (the latter disguised as the Bar Restaurante de Tejeda). These are basic but (mostly) cheap and are much closer to the areas of greatest geological interest. Tejeda also has the advantage of being cooler than places on the coast and having easily the most spectacular morning and evening views of any town on the island.

Weather on Gran Canaria is a happy medium between the eastern and western islands: not too wet in the forest zones in the north, not too hot and dry in the south, although precautions should still be taken against excessive exposure to the sunshine and against dehydration (see Ch. 2). It can be cold and wet in the mountains, and the rugged terrain and many canyons and barrancos mean that special care should be taken during and after heavy rain to avoid being caught out by flash floods.

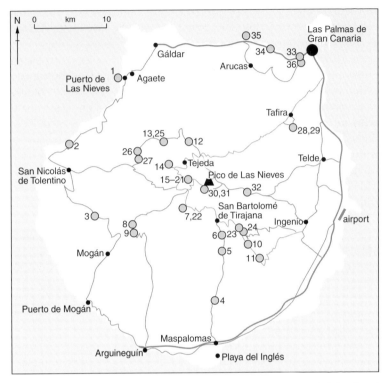

Figure 5.4 Map of locations on Gran Canaria, with major towns also indicated. Only selected roads are shown in the northeast of the island; autopistas indicated by heavier lines.

Shield-stage rocks in Gran Canaria

The best way to obtain an overview of the shield-stage rocks of Gran Canaria, ranging from the early basalts to the later caldera and associated felsic rocks, is in a road tour from Las Palmas, west through Agaete and then south to San Nicolás de Tolentino and Mogán.

The route proceeds north and then west from Las Palmas, along motorway GC-2. Pre-Roque Nublo conglomerates in large roadcuts beside the autopista are capped by nearshore marine sediments, principally carbonate-rich sands, of the Las Palmas formation. These are intercalated with basaltic breccias and basaltic pillow lavas formed when the earliest lava flows of the Roque Nublo Group entered the sea. Subaerial lavas of the same formation cap the cliffs close to Las Palmas, but farther west the Roque Nublo formation almost disappears and most exposures are of more recent Plio-Pleistocene lavas. Pleistocene scoria cones are visible

102

around Guía and Gáldar. South of Gáldar, the young lavas form a narrow coastal fringe at the foot of steep hills and canyons incised into Miocene shield-building rocks.

Stop 5.1 Puerto de las Nieves Turn left just on the south side of Agaete and travel towards the coast at Puerto de Las Nieves. Park in the town and walk along the harbour wall to gain an excellent view of a coastal cliff composed of Miocene basalts, which are greenish as a result of alteration. These are cut by many dykes and inclined intrusions. The top of the cliff is capped by younger valley-filling lavas of Pleistocene age, overlying conglomerates deposited in flash floods.

Beyond Agaete, the route (now the C-810) climbs up spectacular coastal cliffs composed of basalt lavas of the early Miocene shield volcanoes (Hogarzales and Güigüí formations). The rocks are cut by many dykes; these and the general south-to-southeast dips of the lavas south of Agaete indicate that the eruptive centre of these lavas was near Agaete or perhaps off shore: intense late Miocene to Recent coastal erosion (and possibly lateral collapse?) is indicated.

This cliff road is winding and in places very narrow; it is not for the faint hearted or easily distracted, especially on weekends when it can be very busy. It is generally quietest on weekday mornings.

Stop 5.2 Mirador del Balcón Mirador del Balcón, towards the southern end of the cliff road, affords views of the coastal cliff section and of the San Nicolás–Aldea basin to the southeast. This basin is the lower part of the Barranco de Tejeda drainage system. Formerly a deep barranco, it has been filled in by sediments and lavas over time (especially since the Pliocene), as the western side of Gran Canaria has subsided, and is now one of the main agricultural areas in the island because of the abundant groundwater that collects in the basin-fill sequence. Perched terraces and truncated alluvial fans on the south wall of the basin indicate that at one time the sediment-fill sequence was even thicker and is now being re-eroded.

The mountains around the San Nicolás–Aldea basin are mainly composed of Hogarzales and Güigüí formation basalts, but at its eastern end these are truncated by the Miocene caldera ring-fault system and juxtaposed against paler, variously coloured, altered felsic rocks of the caldera-fill sequence, mainly Fataga formation phonolites.

South of San Nicolás, the C-810 follows a concentric barranco (perhaps the only one in Gran Canaria) eroded along the line of a zone of altered rocks at the rim of the Miocene caldera. Dark Miocene basalts are again exposed in the cliffs to the southwest, but to the northeast the cliffs contain a vividly coloured sequence of altered phonolitic (Fataga formation) rocks that underwent intense hydrothermal alteration after emplacement within the Miocene caldera. The alteration has not been studied in detail, but likely contributors to the remarkable range of colours are Fe-rich chlorite, zeolites, very fine-grain haematite and various iron-hydroxide minerals. The caldera wall, locally exposed at the base of the Pajonales cliffs, is not vertical, but dips gently to the northeast; it appears to be a topographic wall, produced by erosion and retreat of the cliff after initial caldera collapse, and the ring fault is not exposed. Its most likely location is to the northwest, beneath the caldera-filling sequence of the Fataga formation.

Stop 5.3 Pajonales cliffs The best place to view and photograph these brightly coloured rocks is at a mirador on the C-810 a few kilometres south of the turn to Tasarte. The mountains they form are part of the Pajonales nature reserve, probably the most remote part of Gran Canaria. Miocene basalts to the west are locally capped by isolated remnants of Pliocene (Roque Nublo Group) lavas, erupted from parasitic vents around the Roque Nublo volcano; it is unlikely that the main stratovolcano extended this far southwest.

The route continues southwards through Mogán. Miocene basalts with many dykes form the walls of the barranco in its upper reaches. The Barranco de Mogán is the southeastern limit of the main exposures of the Hogarzales and Güigüí formations; southwest-dipping felsic lavas of the Mogán formation cap the rim of the barranco around Mogán and form an increasing proportion of the wall sequence farther south.

At Puerto de Mogán the C-810 joins the C-812 and this road is followed along the coast to the east through Puerto Rico and Arguineguín. The rocks exposed in this barren landscape are the extracaldera sequences of the Mogán and (above and to the east) Fataga formations. Most of the rocks along the road are lavas and lava-like rheomorphic ignimbrites of trachytic and rhyolitic composition, but towards the top of the Fataga formation (around Arguineguín) pale weakly welded ignimbrites become more common and are intercalated with massive sheet-like bodies of conglomerate (hyperconcentrated sheet-flood deposits). These are well exposed in the roadcuts along the Arguineguín–Maspalomas section of the autopista. They mark the decline in volcanic activity at the end of the shield-building period and the onset of significant erosion as major drainage systems began to develop on the surface of the volcanoes.

The Miocene caldera

As discussed above, the central region of Gran Canaria is mainly occupied by a Miocene caldera. This is some 15 km across in the southwest–northeast direction and slightly longer in the northwest–southeast direction. It is no longer an obvious topographic feature, because the caldera was filled by Miocene volcanic rocks and intrusions after it formed, deeply eroded after the end of shield-stage volcanism, re-buried by the Pliocene Roque Nublo volcano, and then eroded again. The northern margin is buried under younger rocks, but in the south the erosion of the deep barrancos has exposed many of the structures developed during formation of the caldera and the subsequent caldera-filling extrusive and intrusive activity. Here a mainly road-based excursion through these rocks is described, beginning at Maspalomas. If you travel via the autopista to Maspalomas–Playa del Inglés to start the excursion, the CV-12-1 is not directly accessible from the autopista and it is necessary to leave the motorway, join the old coast road (C-812) where it passes through the north side of the resort and look for a large roundabout that forms the junction of the C-812 and the Fataga road.

From Maspalomas, take the Camino Vecinal 12-1 towards Fataga and San Bartolomé de Tirajana. The road initially climbs a semi-arid hillside formed by the southward-inclined dip slope of the Fataga formation lavas, with the southern end of the Barranco de Fataga to the west. After about 8 km, at a large mirador, the road passes over the lip of the barranco. The mirador offers excellent overviews of the Miocene rocks.

Stop 5.4 Mirador de Achebuche   The mirador is on the eastern side of the barranco, so the best views are in the morning. Below the mirador, the floor and lower walls of the barranco expose a well stratified sequence of strongly welded ignimbrites in the lower part of the Fataga formation; from a distance, these look like lava flows. Above these are true lavas of the Fataga formation. To the north, the eastern side of the barranco is formed by many landslides at the foot of cliffs formed by Fataga formation lava flows. The western side, in contrast, is mainly very well stratified Fataga formation ignimbrites, capped by lavas (Fig. 5.5). The contrast reflects the fact that the trend of the barranco is oblique to the southeasterly dip of the sequence, which results in progressively older rocks being exposed to the northwest.

From the mirador, continue down the road into the barranco via a series of precipitous hairpin bends. Passengers may prefer to fix their gaze on the columnar-jointed intensely welded ignimbrites and lavas, intercalated with sedimentary breccias, on the cliff side of the road.

On reaching the floor of the barranco, continue north towards Arteara.

Figure 5.5 View looking north-west into the Barranco de Fataga from the mirador (stop 5.1), showing the south-dipping well stratified sequence of Miocene Fataga formation ignimbrites and lavas.

On the north side of the village (after a narrow section), park on the right of the road with a view of the western side of the barranco and the cliffs of Lomo del Guardia.

Stop 5.5 Horno del Guardia The cliffs on the western side of the barranco show a marked transition from well stratified rocks to the south to more massive rocks to the north. These form part of the sequence of rocks that accumulated within the Miocene caldera. The actual caldera boundary fault is exposed only in the floor of the barranco and is generally buried by the alluvium that fills its bed.

The caldera-filling sequence is also more heavily altered than the rocks to the south, as can be seen in the many roadcuts farther to the north, where the road climbs up the northern end of the Barranco de Fataga. The road continues over the col to Barranco de Tirajana, the largest barranco in southeast Gran Canaria.

Stop 5.6 Balcón de Tunte Near Balcón de Tunte, about 3 km south of San Bartolomé de Tirajana. The car-park of a large restaurant provides an excellent point from which to view the northeast wall of Barranco de Tirajana. Barranco de Tirajana is another of the very long-lived barrancos in the interior of Gran Canaria. Its south wall is formed by a complete sequence of Miocene felsic rocks, mostly extracaldera rocks of the Fataga formation (phonolitic lavas and lava-like ignimbrites, as in Barranco de Fataga). On the north side a very thick sequence of Roque Nublo Group rocks (including the type section of the Tirajana formation, above the town of Santa Lucía) directly overlies basalts of the oldest Miocene formations (Hogarzales and Güigüí), implying intense pre-Roque Nublo erosion. This suggests that the barranco originally formed somewhat to the north of its present position, but, as the Roque Nublo volcano built up to the north, the barranco migrated southwards. However, the floor of the barranco is mainly formed by recent sediments, including an active land-slip at the foot of the cliffs below Pico de Las Nieves, identifiable by the prominent radio and radar stations, and Plio-Pleistocene lavas.

On the north side of the barranco, the sequence of lavas and ignimbrites dipping and thinning southeast is cut by a prominent phonolite plug, Risco Blanco ("white crag"; note confirmation of the location of the centre of the Roque Nublo volcano, which lies northwest of this location, near Tejeda). The Santa Lucía landslip is beneath this crag; other recent but mainly inactive landslips around the barranco include the ground on which Balcón de Tunte is built; note the very steep faces of Morro de las Vacas and Lomo del Guardia to the west, which are old landslip and rock-fall scars.

Figure 5.6 Barranco de Tirajana viewed from near Balcón de Tunte. San Bartolomé de Tirajana in middle ground, with landslips forming hummocky ground north of town. Cliffs of Barranco de Tirajana behind expose Risco Blanco phonolite plug, at right of view.

The road continues north to San Bartolomé de Tirajana. Turn left at the road junction onto the C-815 (towards Ayacata and Tejeda) and pass through the town, continuing northwestwards. The road winds up slope over recent but stable landslips, below high cliffs composed of Roque Nublo Group rocks (see p. 123 for description of the latter), to Cruz Grande. The road continues for 2.5 km. Turn right onto a side road and continue down hill for 1 km to a car-park at the foot of Morro de Santiago.

Stop 5.7 Morro de Santiago   Morro de Santiago (an excellent picnic site) is formed by a thick sheet intrusion of phonolite, emplaced right at the end of the Miocene (shield-stage) volcanic activity. The intrusion, up to 50 m thick, dips steeply southwards. It is emplaced into caldera-filling volcanic rocks and older intrusions of phonolitic rock, mostly north-dipping inclined sheets that form the southern sector of the cone-sheet swarm centred on the Tejeda area (Fig. 5.2). These rocks form most of the lower walls of the barrancos to the south and west, and also much of the high peaks to south and west as far as Montaña de Tauro. This distinctive flat-top peak lies outside the caldera rim; it is capped by well stratified post-caldera (Fataga formation) volcanic rocks. Within the barranco, the intervening low plateaux to the south are formed by the remnants of Pliocene rocks that once filled the barranco, and which also form the spectacular cliffs to the north, leading up to Roque Nublo itself; these are discussed below.

Return to the main road and continue north to Ayacata (with a couple of restaurants, only open during the day, as alternative sites for lunch). From here there are two alternative routes, one south and west and the other back to the east. The choice of which to follow depends in part on where you are based: go south to the south coast resorts, or east to join the autopista and end up either in Las Palmas or in Maspalomas. However, the southwest route is partly weather dependent, because of a poorly surfaced road. It is also possible to continue north to Tejeda to examine the centre of the Miocene caldera, especially if staying there. These localities are described separately below (pp. 127–131).

Southwest route

Just west of Ayacata, bear left and down hill onto a side road (C-811, for Soria). The route is initially through rockfall and landslip deposits from the Roque Nublo Group rocks in the cliffs above, but the *in situ* rocks farther to the west are heavily jointed inclined sheet intrusions of the Miocene post-caldera intrusive complex. Originally of phonolitic to trachytic composition but now heavily altered, they form an annular cone-sheet swarm around the centre of the intrusive complex, near Tejeda. The out-crops here are on the south side; the intrusions dip uniformly to the north.

After about 3 km the terrain levels out onto the irregular top of a plateau between three barrancos: Arguineguín to the south and east, Mogán to the west and (beyond the pine-covered ridge of Morro de Pajonales – often closed because of fire risks) the Barranco de Juncal to the north. The latter is a tributary of the Barranco de Tejeda. The southern part of the plateau is covered by Roque Nublo debris-avalanche breccias, locally extending up to the road, but most exposures in cliffs and roadcuts are of Miocene caldera-filling rocks. These gradually change to the south and west, with the proportion of inclined sheet intrusions decreasing and the proportion of flat-lying to south-dipping volcanic rocks (mainly altered lavas, so it is not always easy to distinguish the intrusive and extrusive rocks close up; look for the contacts between rock units instead). The intensity of altera-tion also decreases, reflecting the approach to the margin of the caldera. Take care if you stop by the side of the road to examine the outcrops; it is better to stop at the picnic area at Embalse Cueva de las Niñas and examine outcrops to the north on foot.

At about 12 km from Ayacata, the road turns south along the col above the head of Barranco de Mogán and turns into an unsurfaced track (caution: this track may be impassable in wet weather).

Stop 5.8 Cruz de San Antonio About 15 km from Ayacata, the col between barrancos de Arguineguín and de Mogán offers an excellent view of the margin of the Miocene caldera, on the western side of Barranco de Mogán (Fig. 5.7). It is defined by a steeply northeast-dipping contact between paler-weathering poorly stratified rocks to the northeast of the caldera-fill-ing sequence (mainly massive ingnimbrites and lavas, with interbedded sediments and breccias), and gently southwest-dipping basaltic lavas to the southwest of the contact. The contact therefore represents the caldera wall, which has been buried by the filling of the caldera. The scale of the structure can be appreciated by the fact that neither the top nor the base of the caldera-filling sequence is exposed, implying that the wall of the cal-dera was a cliff well over 0.5 km high.

Continue southwards along the track. A junction about 2 km south of Cruz de San Antonio marks the start of a steep and difficult unsurfaced track that descends into Barranco de Mogán; bear south along the rim of the barranco instead and return onto a tarmac surface. The outcrops along the road for the next couple of kilometres are actually of Roque Nublo (Pliocene) debris-avalanche breccia that spilled over the rim of Barranco de Arguineguín during the third collapse of the Roque Nublo volcano.

Figure 5.7 Topographic margin of Miocene caldera, exposed in the western wall of Barranco de Mogán, as viewed from Cruz de San Antonio. Steep, northeast-dipping contact separates flat-lying Miocene basaltic lavas from east-dipping felsic ignimbrites and lavas.

Stop 5.9 Montaña de Tauro At the southern end of the col, the road bears left and descends into Barranco de Arguineguín. Just before it does so, you cross over the rim of the Miocene caldera and there is a sharp transition to more clearly stratified rocks, best exposed in the cliffs of Montaña de Tauro, which towers above the western rim of the barranco. The energetic may wish to backtrack and walk up a track that ascends the northern ridge of the mountain; others should park and examine the lower walls of Barranco de Arguineguín and, with care, exposures along the road. The rocks along the road are extracaldera phonolitic ignimbrites (very strongly welded, usually flow banded and difficult to distinguish from lavas; look for the remnants of streaked-out pumices or **fiammé** and occurrences of **lithic fragments**). These are contemporaneous with the formation of the caldera. They pass up hill in Montaña de Tauro into a postcaldera sequence – the erupted equivalents of the intracaldera intrusive complex – of ignimbrites and true lavas.

Descend to Soria and take the road down the barranco to Arguineguín. The same sequences of syn-caldera and post-caldera rocks occur in the higher walls of the barranco, whereas intermittently exposed at road level are the older basaltic lavas that underlie them.

East route

Return to San Bartolomé de Tirajana and continue through Santa Lucía on the C-815 (signposted to Aguimes and Vecendario), stopping 5 km southeast of Santa Lucía at a small viewpoint overlooking Fortaleza Grande, one of two pinnacles of lava flows within the barranco.

Stop 5.10 Barranco de Tirajana overlook   These pinnacles are remnants of a Plio-Pleistocene sequence of lavas that infilled Barranco de Tirajana about 2 million years ago and have since been eroded away as the barranco was re-incised; the extensive areas of alluvium in the floor of the barranco suggest that at present (since sea level is high) the barranco is filling again. On either side of the floor of the barranco, this recent erosion has exposed a sequence through Miocene rocks, which is cut by the southeastern limit of the Miocene caldera (Fig. 5.2). This is best seen in the cliffs below the two peaks of Amurga and Garita to the southwest. The tops of these mountains (*c*. 1100 m elevation) are formed by post-caldera lavas and ignimbrites of the Fataga formation, but at about the level of the prominent track running southeast from the village of Sitio de Abajo, some 200 m above the floor of the barranco, the sequence is split by a prominent contact, near-vertical at the base of the cliff and curving to an apparent northwesterly dip higher up. To the southeast of this, the lowest section of cliff is formed by a sequence of thin basaltic lavas, part of the early Miocene shield volcanoes, with a *c*. 100 m-thick single lava flow above (pale weathering and rhyolitic in composition). The same rock type occurs to the northwest of the contact but forms the whole of the lower cliff. Its base is not exposed and it is much thicker, as here it is infilling the caldera. The increase in thickness coupled with the vertical fault contact suggests that the caldera formed progressively, with multiple episodes of subsidence accompanied by progressive infilling with thick lavas (as here) and pyroclastic flows. The shallower contact (probably the topographic wall of the caldera) at the upper part of the lower cliff, just below the road, coupled with draping by thinner-bedded ignimbrites, suggests the final stage of infilling of the caldera after movement on the fault had ceased.

Continue southeastwards around a series of sharp bends cut into Roque Nublo Group ignimbrite breccias and later lava flows, for about 3 km, to Montaña de Las Carboneras. A large parking bay offers further views over the lower course of Barranco de Tirajana, where it is incised along the line of the ancient barranco and therefore separates Miocene rocks on the southwest side from sequences dominated by Pliocene through to Pleistocene lavas on the northeast.

Stop 5.11 Montaña de Las Carboneras   Montaña de Las Carboneras itself is composed of an inlier of Miocene rocks, and although at the southeastern limit of the exposure of these rocks the road cuts back towards Santa Lucía (caution is required while walking this section, as the road is narrow and often busy) offer exceptional exposures of the characteristic rock type of the Miocene sequence in Gran Canaria, highly welded rheomorphic or lava-like ignimbrites. These are strongly flow-banded rocks with only ghosts of fiammé remaining, but close examination reveals these and also the frequent presence of lithic fragments around which the flow banding and fiammé are deformed. Another distinctive feature of these rocks is the intense folding of the flow banding, defining southeast-directed recumbent and sheath folds that result in the strange elliptical and circular patterns of flow banding seen on southeast-facing surfaces in particular. This took place after deposition when the ignimbrite compacted into a very hot mass, capable of flowing down hill like a viscous lava. The unusually high deposit temperatures (several hundred degrees) required for this to occur imply rapid deposition of the ignimbrite from a hot and dense eruption cloud that cooled only slightly between eruption and deposition.

After returning to the parking area, continue along the C-815 to Aguimes or turn right about 2 km farther on (after rounding the tip of Montaña de Las Carboneras) to Sardina del Sur, and join the autopista at either place. Sardina del Sur and the surrounding agricultural land occupy the large alluvial fan, some 10 km across and hundreds of metres thick, that has formed at the mouth of Barranco de Tirajana as it has been re-incised.

Pliocene rocks in central Gran Canaria
The Roque Nublo volcano and lateral collapse structures

This excursion considers the Pliocene post-erosional rocks of the Roque Nublo volcano in the vicinity of its central crater complex around Tejeda (Fig. 5.2). The itinerary assumes a start in Las Palmas; Tejeda itself is also a good base for the excursion (in which case drive towards Las Palmas on the C-811 as far as Cruz de Tejeda, and join the itinerary there).

Leave Las Palmas in the direction of Tejeda, on the C-811 through Tafira. Roadcuts along the route from Las Palmas to Tafira expose alluvial fan conglomerates containing green phonolite lava and welded ignimbrite clasts of the Fataga formation; these deposits were formed by pre-Roque Nublo erosion of the Miocene volcano.

The northeast of Gran Canaria is heavily vegetated and cultivated, and exposures of bedrock are mainly confined to the beds of the barrancos as far as San Mateo (22 km). To the west of San Mateo, the ridges to the north

113

and south are mainly formed by Roque Nublo Group rocks, with later Plio-Pleistocene lavas (Llanos de La Pez formation) filling the valley floor and forming the ridge to the west. These latter lavas, with intercalated baked red lapilli and soil beds, form a high crag at the turn-off of the CV-15-6 a few kilometres short of Cruz de Tejeda.

The route along the C-811 continues to the Parador de Cruz de Tejeda (37 km from Las Palmas); here a right turn through the car-park leads to the road to Pinos de Gáldar. After 3.5 km, a mirador at Degollada de Las Palomas gives a spectacular view of the Barranco de Tejeda (Fig. 5.3).

Stop 5.12 Degollada de Las Palomas (Fig. 5.8) The Barranco de Tejeda, over 8 km wide and in places over 1 km deep, is by far the largest erosional barranco on Gran Canaria. It has existed in roughly its present position for over 5 million years; the Roque Nublo volcano filled it up at least twice and it was eroded out again in less than a million years. At the present day, the floor of this immense canyon is mainly formed by Miocene rocks of the intracaldera fill (Fataga formation) and intrusions emplaced into this. The far southern rim of the barranco is also formed by Miocene rocks; stratified rocks in the far distance to the southwest are the sequence of Miocene lavas and ignimbrites outside the caldera. Various mesas and cumbres within the Barranco de Tejeda, and its northern and eastern walls, are formed by Roque Nublo Group rocks and later Plio-Pleistocene rocks.

The immense northern wall on which the mirador is situated, the Roque Bentaiga ridge in the middle and the Montaña del Humo ridge to the south are composed of early Roque Nublo Group lavas and ignimbrites, as are parts of Mesa de Junquillo and Mesa de Acusa, within the valley to the southwest (here the valley-filling nature of the Roque Nublo Group can be seen most clearly). The upper parts of Mesas de Junquillo and Acusa are however formed of younger Roque Nublo Group rocks (debris-avalanche breccias of the third collapse) and post-Roque Nublo Group lavas and sediments, respectively; these bear testimony to the re-excavation of the Barranco de Tejeda during the life of the Roque Nublo volcano. To the south, the pinnacle of Roque Nublo and the massive cliffs below Montaña del Aserrador and El Montañón are formed by the lateral collapse breccias and slide blocks (see stop 5.14 and also pp. 118–123). The original summit region of the Roque Nublo volcano lay between the northern wall of Barranco de Tejeda and Roque Nublo, and consisted of a north-west–southeast elongated crater complex. Most of the rocks emplaced in this crater were soft because of intense hydrothermal alteration and have been eroded away. As a result the crater area is now occupied by the late Pliocene Llanos de La Pez lavas forming the cliff above Tejeda (also previously seen on the road up to Cruz de Tejeda from the east). However, the

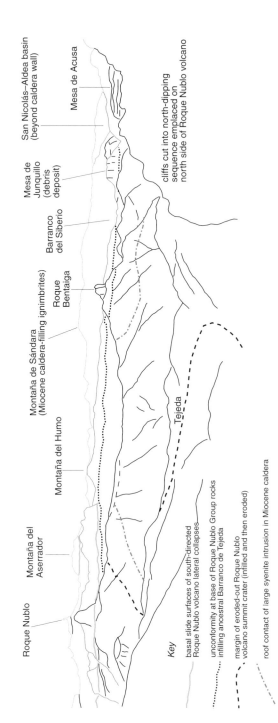

Roque Nublo

Montaña del
Aserrador

Montaña del Humo

Montaña de Sándara
(Miocene caldera-filling ignimbrites)

Roque
Bentaiga

Tejeda

Barranco
del Siberio

Mesa de
Junquillo
(debris
deposit)

San Nicolás–Aldea basin
(beyond caldera wall)

Mesa de Acusa

cliffs cut into north-dipping
sequence emplaced on
north side of Roque Nublo volcano

Key

.... basal slide surfaces of south-directed
Roque Nublo volcano lateral collapses

unconformity at base of Roque Nublo Group rocks
infilling ancestral Barranco de Tejeda

– – – margin of eroded-out Roque Nublo
volcano summit crater (infilled and then eroded)

–·–·– roof contact of large syenite intrusion in Miocene caldera

Figure 5.8 Sketch of the view into the Barranco de Tejeda from Degollada de Las Palomas on its northeastern rim, with major contacts within the Miocene caldera fill and intrusive complex and the unconformity at the base of the Pliocene Roque Nublo Group indicated. Note that the latter is very complex, indicating that the Roque Nublo volcano infilled the ancestral Barranco de Tejeda.

western margin of the summit crater complex can be traced across the floor of Barranco de Tejeda, west of the town, as a transition between poorly exposed rocks around the town and the reddish rocks of the Miocene intrusive complex.

Continue along the road at the top of the ridge, through pine forests and alpine meadow vegetation, to Mirador de los Pinos de Gáldar; turn left there towards Artenara.

Stop 5.13 Overview of Barranco de Tejeda from Artenara   After parking in Artenara, walk past the church to the terrace of a café that also overlooks Barranco de Tejeda but offers clearer views of Mesa de Acusa and Mesa de Junquillo, the erosional outliers of Roque Nublo Group rocks. The rocks forming these mesas are considered in more detail below (pp. 127–131, lower Barranco de Tejeda itinerary). The much drier conditions, abundance of bare rock and sparser semi-arid vegetation of the south of the island become apparent from here onwards.

Rejoin the main road, and at the western end of Artenara turn back onto the CV-17-5 towards Tejeda. This road winds along the foot of the cliffs forming the north wall of the Barranco de Tejeda; it cuts down to the east through the lower part of the Roque Nublo Group as far as El Rincón, which is just 1 km horizontally (but 500 m vertically) from Degollada de Las Palomas. Cliff exposures of Roque Nublo Group lavas and breccias, intruded by northwest-trending dykes, are visible on one side; on the other, the red (the result of alteration) Miocene intrusive rocks of the Tejeda "formation" form the valley floor below. The road then climbs back up to rejoin the main (C-811) road at Tejeda, a picturesque town straggling along the hillside.

Just beyond Tejeda, the C-811 crosses the floor of the upper Barranco de Tejeda; most of the roadcuts are in Miocene rocks, but the valley floor is partly filled by a Pleistocene lava flow with excellent columnar jointing, and Roque Nublo Group rocks are visible in the crags to the east; these intensely altered rocks are part of the early crater-fill sequence. The bend at the floor of the barranco marks the start of a path to Roque Nublo that is one of the most interesting walks in the area.

Five kilometres from Tejeda, the road climbs back up through early Pliocene basaltic lavas of the Riscos de Chapín formation. At the crest of the ridge, turn off towards the very prominent Roque de Bentaiga and then turn left about 0.5 km farther on to follow the new road to the Cuevas del Rey visitor centre car-park. The Cuevas del Rey are Guanche caves perched high on the southern side of Roque Bentaiga and are worth a visit if you are archaeologically inclined.

Stop 5.14 Roque de Bentaiga parking area   From the parking area, the cliffs below Degollada de Las Palomas and Artenara can be seen to the north. A complete *in situ* sequence through the Riscos de Chapín and Tirajana formations of the Roque Nublo Group can be seen. The lavas and breccias thin and dip westwards along the cliff; their true dip and thinning direction is to the northwest, but the northward component is not visible from this viewpoint. Measurement of the dip directions and angles has enabled calculation of the position and height (2500–3000 m) of the summit of the Roque Nublo volcano to the southeast. Note also that the dykes cutting the sequence are not perpendicular to the cliff but trend northwest–southeast, also helping to locate the original position of the summit of the volcano.

To the east is a spectacular view of the sheer cliffs below Roque Nublo itself and Montaña del Aserrador. Roque Nublo, the remarkable pinnacle at the northern end of the cliff, is discussed below. This area contains the remnants of no fewer than three lateral collapses of the Roque Nublo volcano; subsequent erosion has provided slices through the lower parts of the collapse deposits and exposures of the basal surfaces upon which the displaced rock masses moved.

The road along the foot of the cliff roughly coincides with the irregular unconformity at the base of the Roque Nublo Group. Below it, altered Miocene rocks of the intracaldera intrusive complex form yellow to rust-coloured crags around the village of Timagada. Poorly exposed *in situ* lavas of the Roque Nublo Group (Riscos de Chapín formation) form most of the sequence between the road and the foot of the cliff. A prominent notch and overhang at the base of the main cliff, with apparent dip to the south (true dip is to the southeast) marks the basal detachment fault of the second Roque Nublo collapse structure (Fig. 5.9). Everything above this fault, and all the rocks to the west of the cliff which have been eroded away, slid southeast and then south-southwest during the collapse, by a distance of several kilometres.

The cliffs, up to 400 m high, consist mainly of a massive breccia sheet that formed as an immensely thick but slow-moving debris flow during the first collapse to affect the Roque Nublo volcano. At the very top of the cliff, the pedestal of the Roque Nublo pinnacle, El Montañón to the north and Montaña del Aserrador to the south are formed by post-second-collapse rocks; and a near-horizontal fault at the base of the pinnacle and near the top of Montaña del Aserrador forms the base of the third collapse structure. These isolated crags are almost all that remains of the sequence of rocks that slid west by several kilometres in the third lateral collapse.

The route continues east along a ridge of Miocene rocks with many west-dipping inclined sheets (it is not clear if the host rocks are earlier sheet intrusions or lavas, because of the intense alteration), then south

117

Figure 5.9 View from below the parking area at Roque Bentaiga, showing the apparently south-dipping basal contact of the second collapse structure to affect the Roque Nublo volcano (the prominent overhang at the base of the cliff).

along the C-811 to Ayacata. The road passes along the foot of the cliffs examined from stop 5.14, then turns east between Montaña del Aserrador and other crags composed of the massive, pitted-weathering breccia produced by the first collapse of the volcano. The dark irregular masses of rock within the breccia are basaltic intrusions emplaced into it shortly after the first collapse when it was still water saturated and fluid. Ayacata is another tiny village (but with plenty of outdoor restaurants), which sits between cliffs formed by the first collapse breccia sheet and irregular intrusions emplaced into it after the first collapse but before the second; these form the Las Candelillas complex, which was emplaced beneath the summit of the Roque Nublo volcano before the entire mass of rock forming these cliffs slid about 5 km to the south-southeast in the second collapse.

Return to Ayacata and turn up hill onto the CV-17-6 (towards Roque Nublo and Llano de la Pez). The road winds up between the cliffs to the head of Barranco de Ayacata. Park in a small car-park at the col (park at the side of the road on either side on busy days, especially at weekends).

Walk around Roque Nublo

This walk starts at the car-park at the col between Barranco de Tejeda and Barranco de Ayacata, and takes about four hours (with stops, including a stop for lunch). It provides interesting insights into the re-growth of the Roque Nublo volcano after the second lateral collapse, contemporaneous erosion, and the subsequent third collapse of the volcano.

Caution: This walk is in places adjacent to the tops of very high cliffs. Do not approach them too closely. The pinnacles of Roque Nublo and La Rana are regarded as sacred and are strictly protected; in spite of this, climbers may create an additional hazard by dislodging rocks, so it is dangerous to stand underneath the pinnacles.

Stop 5.15 Col between Barranco de Tejeda and Barranco de Ayacata   The path west of the car-park cuts through a porphyry intrusion breccia. Much of the rock has undergone high-temperature **fumarolic** alteration, but the larger blocks contain relict **flow banding**. A prominent **phenocryst** and **megacryst** assemblage of feldspar, pyroxene, amphibole (plus rare biotite and sphene) is present, as are rare small **microsyenite** blocks or xenoliths. The intense brecciation is typical of all the intrusions within the summit area of the Roque Nublo volcano. It seems to have been caused by many small explosions as magmatic gases and steam produced by heating of groundwater escaped up through the intrusions as they cooled; it is likely that these intrusions fed volcanic vents at the surface, from which rocks similar to the distinctive Roque Nublo ignimbrites were erupted (stops 5.17, 5.18 and 5.20; see also p. 124 (Tirajana) and pp. 129–129 (Mesa de Acusa)).

Stop 5.16 Roque de Aguja   Exposures on the left of the path as it climbs up to the strange pinnacle of Roque Aguja are of mixed (or **polymict**) breccias, superficially similar to the Pargana breccias but with a clast assemblage that includes rock types that postdate the second collapse (however, these are rare and they are seen *in situ* at later stops; they will be considered again later), implying that they were deposited during re-growth of the volcano between the second and third collapses.

At the base of Roque Aguja ("needle"), the main path continues up slope, but the path to the next stops forks to the right, slightly down slope, and passes along the north side of the plateau on which the immense pinnacle of Roque Nublo sits (if time is limited, skip stops 5.17–19 and proceed up slope to stop 5.20 at the eastern end of the platform on which the pinnacle sits).

Stop 5.17 Path north of Roque Nublo   Shortly before the path reaches the col between Roque Nublo and El Montañón, fork left and slightly down slope to a low east-facing crag among trees. This is composed of beds of a breccia with a tuffaceous matrix, green altered **pumices** (especially common at the tops of certain beds), and very abundant clasts of older rocks, including **microgabbros** and **porphyries**. The latter two rock types match rocks in the post-first-collapse intrusive complexes; thus, these rocks are part of the post-second-collapse sequence, as also seen at stop 5.16. The origin of this type of breccia, which is very common in the Roque Nublo Group (see stop 5.24, below, p. 124), has been something of a puzzle, but it is now interpreted as lithic-rich ignimbrite, as discussed at other outcrops of these rocks (pp. 129–129).

Stop 5.18 El Montañón (optional, depending on time and weather)   The ridge to the north of stop 5.17, terminating in a prominent crag of El Montañón, is composed of a sequence of soft, sedimentary or **epiclastic** breccias and lavas, also part of the post-second collapse sequence. The most distinctive unit in this sequence is a lava flow with large platy plagioclase-feldspar phenocrysts; plagioclase usually forms elongate lath-shaped phenocrysts, but in highly alkaline basanites and **tephrites**, such as occur in the Canaries, plagioclase plates are common. The crag to the north (El Montañón itself) is composed of strongly indurated lithic-rich ignimbrites.

To the west are fine views of the Barranco de Tejeda and westernmost Gran Canaria. To the east, the cliff that forms the headwall of Barranco de Tejeda, above the village of La Culata, exposes intercalated lavas and sediments (mainly lacustrine in origin) of the Llanos de La Pez formation. A section through a maar-like crater can be seen in the cliff north of La Culata. The Llanos de La Pez formation unconformably overlies rocks of the Roque Nublo Group; these are mainly altered intrusive rocks and lavas and sediments of the crater-fill sequence (see stop 5.12), but a prominent crag above La Culata is composed of massive, Pargana member, polymict breccia and brecciated porphyry intrusions. The base of the crag is formed by the same fault that underlies the Roque Nublo Massif, the fault that formed the detachment surface during the second collapse (see stop 5.14).

The route returns to the path around the base of the plateau on which Roque Nublo sits and continues southwest. A few tens of metres on, it emerges from the trees, and good views can be gained of the pinnacle and plateau (Fig. 5.10).

Stop 5.19 Path northwest of Roque Nublo   The plateau beneath Roque Nublo consists of strongly indurated ignimbrites and felsic spatter units (see stop 5.20), but on the north side of the plateau these overlie soft pale ignimbrites that fill another small palaeo-barranco developed between the second and third collapses, again indicating active erosion that punctuated the growth of the volcano.

The path continues south to a junction; turn east through another wood of Canary pines, with views through the trees of Montaña del Aserrador and other crags to the south. The prominent notch along the face of Montaña del Aserrador is formed by the detachment fault of the third collapse

Figure 5.10   View of the Roque Nublo pinnacle from the north, between stops 5.18 and 5.19. The platform on which it sits consists of *in situ* ignimbrites emplaced between the second and third collapse of the volcano, but the pinnacle itself is one of the last remnants of a thick slide mass transported some kilometres from the east in the third collapse. The notch at the base of the pinnacle is eroded in the sheet of soft mud on which this slide mass moved (see stop 5.21).

structure to affect the Roque Nublo volcano; the rocks above are a tiny remnant of a mass that slid several kilometres west during this collapse.

The path continues eastwards and climbs up to the col between Roque Aguja and the Roque Nublo plateau. Turn west up onto the plateau. This is formed by a sequence of volcanic rocks emplaced during the post-second-collapse re-growth of the volcano.

Stop 5.20 Steps up to Roque Nublo Plateau   The two thickest units of this sequence both show a distinctive gradation, from an indurated but unwelded and relatively fine-grain lithic-rich ignimbrite at the base of the unit, upwards through a coarser pumiceous component into a striking rock containing flattened phonolitic spatter clasts (Fig. 5.11); these have dark glassy porphyritic cores, with abundant feldspar, pyroxene and amphibole phenocrysts. It seems that the eruptions that produced these beds changed from an initial Vulcanian phase, which generated the unwelded

121

Figure 5.11 Unwelded ignimbrite grading up into rock containing dark phonolite spatter clasts, beside the path at the eastern end of the Roque Nublo platform. The notebook 15 cm wide.

ignimbrite at the base of each unit, into vigorous fire fountains that deposited these spatter beds near the vent. The initial explosively erupted but relatively cool deposit, with its lack of welding, suggests that the initial explosions were hydromagmatic in origin (i.e. driven by the explosive expansion of heated groundwater). A distinctive feature of these units, and especially of that which forms the main plateau, is an abundance of coarse-grain pyroxenite xenoliths; these seem to be crystal cumulates that formed on the walls and floor of the magma chamber in which the phonolitic magmas collected before being ripped up and erupted with the magma.

The plateau on which Roque Nublo and its smaller counterpart, La Rana ("the frog"), stand was once a Guanche holy place in which the worship of the sky gods was conducted. It is currently a protected site, so no collecting is allowed.

Stop 5.21 Roque Nublo and La Rana The two pinnacles, Roque Nublo and La Rana, are unrelated to the rocks beneath them. The contact relationships between them and the rocks beneath are best seen on the south side of La Rana.

The path around Roque Nublo itself is very hazardous, even when climbers are not knocking rocks down onto it. Do not attempt it.

The overhang at the base of La Rana has formed by the erosion of a soft weathered-out sheet of streaky banded clayey rock, containing altered pumices. Complex flow folds can be seen in the more distinctly banded parts of the rock. This is a **fault gouge**, a soft mud-rich rock that was crushed and which flowed during slip on a fault. The presence or absence of a fault gouge, particularly one that was highly pressurized as this one was (note the veins of gouge injected into the roof rocks), is critical to the behaviour of a fault because, if fluid fault gouge is present, the rocks above the fault effectively float on it and can slide great distances very easily, on almost horizontal fault planes. The overhang itself (especially on the roof of a miniature cave a few metres west of the eastern end of the rock, on the south side) is grooved, with very fine west-southwest directed grooves that indicate the direction of movement. Together with the age of the rocks beneath the fault (which are known to postdate the second collapse), this indicates that this fault is part of the third collapse structure.

It is important to remember that Roque Nublo, La Rana, Montaña del Aserrador and the other remaining crags above the detachment fault are but tiny remnants of a much thicker continuous mass of rock that slid several kilometres to the west during the third collapse of the Roque Nublo volcano. This collapse produced a debris-avalanche deposit that has also been mostly eroded away.

The return route to the car-park to the east involves going back to the eastern end of the plateau (stop 5.20), turning northeast and walking down past Roque Aguja. On returning to the road, it is possible either to continue east through Llano de la Pez to Cruz de Tejeda and thence either to Tejeda or Las Palmas, or if time permits to continue as follows.

Return to Ayacata and turn left onto the C-815 (to La Plata and San Bartolomé de Tirajana). Continue south around the foot of spectacular cliffs composed of Pargana breccias and intrusive rocks.

Stop 5.22 Morro de Santiago   The cliffs can be best viewed from Morro de Santiago, the Miocene intrusion mentioned above (p. 109; turn right at junction 3 km south of Ayacata). The massive breccias within the collapse structure form the cliffs above the road almost as far as Cruz Grande, with only a thin discontinuous layer of Roque Nublo Group lavas separating them from altered Miocene rocks below. This is because the basal fault of the collapse cut deep into the volcano and removed most of it. At the eastern end of the cliff above La Plata, the massive breccias have a vertical faulted contact against thick-bedded *in situ* breccias, above a thin sequence of basaltic lavas; this is the La Plata fault, which forms the eastern side of the second collapse structure.

To the south and west of Morro de Santiago the mesas and cumbres (with undulating flat tops) within Barranco de Arguineguín are composed primarily of debris-avalanche deposits emplaced down slope from the collapse structures.

Return to the main road and continue east towards San Bartolomé de

Tirajana. On the eastern side of the Cruz Grande ridge, the road emerges at the top of Barranco de Tirajana. The north side of the Barranco is formed by a very thick *in situ* sequence of Roque Nublo Group rocks. The floor of the barranco is however mostly formed by recent sediments and Plio-Pleistocene lavas; these and bodies of Roque Nublo Group rocks are involved in some large creeping landslips extending from above San Bartolomé de Tirajana to Santa Lucía.

Continue down hill on the C-815 through San Bartolomé to Santa Lucía. Much of the ground over which the road passes is occupied by relatively inactive landslips, but at the floor of the barranco (at the entrance to Santa Lucía) continuing erosion at the toe of the landslips on the north side of the barranco results in movement.

Stop 5.23 Santa Lucía bridge   Park in the village and walk back to the barranco. The current bridge across the barranco is relatively recent (built in the 1950s). Just down stream are the remains of an older bridge, the centre of which has collapsed. Examination of the two halves (look along the line of the walls) reveals that the northern side is displaced to the east of the southern half as a result of downslope movement of the landslide on the northern side. Much of this movement occurred in a single storm in 1952, reflecting the sensitivity of landslip movement to groundwater conditions. **Do not venture onto the bridge, which remains highly unstable.**

Continue east through Santa Lucía on the C-815. One kilometre east of the town, bear left on the CV-2-1 towards Aguimes. Park on the left a few hundred metres up this road.

Stop 5.24 Santa Lucía–Aguimes Road   Continue up hill on this road for about 200 m, to prominent roadcuts where thick-bedded southeast-dipping rocks descend to meet the road level. These are lithic-rich breccias, with many angular clasts of various lavas and scoriaceous rocks, but they also contain pale-green pumice blocks and an abundant pumiceous matrix. Also present are the cavities or moulds left by the decay of tree trunks and branches that were incorporated into the deposits. These are the type examples of the so-called Roque Nublo ignimbrites, also seen at stops 5.17 and 5.18. Their origins have been much debated, but it seems that they represent the low-temperature end of the range of ignimbrite deposits, having been produced by hydrovolcanic or steam explosions at shallow depths (note the absence of Miocene rock types in the clast assemblage, implying that the lithic fragments were derived from the vent walls within the Pliocene sequence) in the summit-crater complex of the Roque Nublo volcano. As such, they make an interesting contrast with the very high-temperature lava-like ignimbrites of the Miocene volcanic activity.

Continue towards Aguimes on the CV-2-1. The road climbs out of the Barranco de Tirajana (providing good opportunities for viewing its south side) and winds north around crags of post-Roque Nublo, Pliocene to Pleistocene lavas, above a highly irregular unconformity. The rocks below this are Miocene in age. Around Temisas, they are pale felsic lavas and ignimbrites, representing part of the later Miocene caldera-fill sequence, but farther down slope to the east beyond a steeply northwest-dipping contact are earlier basaltic lavas. The contact is not well exposed but it represents the extreme southeast limit of the Miocene caldera: it will be noted that this extends almost two thirds of the way across the island in this northwest–southeast direction.

Bear right in Aguimes towards Cruce de Arinaga, on the coastal plain, and join the autopista for the return journey to Las Palmas or Maspalomas.

Geological walks in the Tejeda area

The upper parts of the Barranco de Tejeda and the surrounding pine-forested cumbres offer some of the best walking in the Canary Islands, especially if Tejeda is used as a base and it is possible to start walking early in the morning.[*]

Tejeda–La Culata–Roque Nublo–Tejeda

The following itinerary takes a full day, especially if some of the alternative routes are followed. Walk south out of Tejeda on the C-815 to the floor of Barranco de Tejeda, where a signposted path now leads up the barranco to La Culata. At this point the barranco is incised into heavily altered Miocene intrusive rocks, which are locally draped by recent lava flows that descended the barranco, but to the east up the barranco the path crosses into well bedded sediments, pumice beds and scoria beds, cut by many basaltic dykes. These are of Pliocene age and are part of the sequence filling the early Roque Nublo crater complex. The well bedded nature of the rocks suggests that at time this crater was filled by a lake through which the pumices and sediments settled. Join the Tejeda–La Culata road and walk into La Culata, before crossing the barranco and climbing up through pine forests on the south side of the barranco.

Rochford's walk 1 can then be followed up to the car-park at the col between Barranco de Tejeda and Barranco de Ayacata, but the rock exposures along this route are limited. Alternatively, about 100 m above La Culata, at a junction in the path level with abandoned buildings to the

---

[*] See Noel Rochford's *Landscapes of Gran Canaria* for some suggestions (see p. 284).

west, turn right onto a path that leads past these buildings. They are built into cliffs (above the top of the crater-fill sediments) composed of thick-bedded breccias similar to the massive debris deposits formed in the first collapse of the volcano, suggesting that similar smaller landslides preceded this event. As the path climbs up past these deposits (note also the prominent northeast–southwest trending basaltic dykes cutting them), take the opportunity to look back north and east to the eastern wall of Barranco de Tejeda, in which many intrusions cutting soft-grey heavily altered rocks are exposed; this is the core of the Roque Nublo volcano. Above these rocks, forming the upper part of the cliffs, is a spectacular sequence, some hundreds of metres thick, of interbedded sediments (including ash beds) and thick columnar-jointed lava flows; this is the Llano de la Pez formation, which infilled a depression eroded into the soft rocks after the third collapse of the Roque Nublo volcano.

Continue up the path. The route takes you up through the level of the basal detachment of the second collapse of the Roque Nublo volcano, which can be traced as a south-dipping fault surface through the cliffs of a barranco to the west of the path, and then up to the level of the Roque Nublo platform. The path joins the well travelled route around Roque Nublo itself (stops 5.17–21, pp. 119–123); by returning along this path to the car-park you regain the route back to La Culata.

From the junction south of stop 5.19, it is possible to descend into the valley south of Roque Nublo, over water-washed surfaces that mark a time when this was a major barranco; it has been abandoned as its headwaters were captured by the growth of the Barranco de Tejeda and Barranco del Chorrillo to north and west, respectively. The crags on either side of the valley – Montaña del Aserrador to the west and Roque Aguja to the northeast – are formed by remnants of the third collapse landslide. If you are used to scrambling, it is worth climbing up either to the prominent overhang east side of Montaña del Aserrador or passing around the southern end of the long ridge south of Roque Aguja to its eastern side, where a similar overhang is present. Both overhangs were formed by erosion of soft fault-gouge muds that formed on the basal fault of the collapse. As a result of the erosion of this material, the extraordinary grooved and polished underside of the landslide mass, scratched by its passage over blocks in the underlying rocks, can be examined in the roof of the overhangs; the grooves give the direction of motion of the landslide to the west-southwest.

Retrace your steps to rejoin the path in the valley: in particular, **do not attempt to scramble along the east side of Roque Aguja to join the Roque Nublo path to the north.**

Barranco de Tejeda
Miocene and Roque Nublo Group rocks

This excursion explores the lower reaches of the spectacular Barranco de Tejeda – arguably the most spectacular canyon anywhere in the Canaries, and one of the most spectacular in any part of Spain – including both the Miocene rocks that form the floor and upper walls of the barranco and the outliers of Pliocene rocks that represent the remnants of the sequences that have more or less filled the canyon at different times in its history.

The description here begins in Artenara, which can best be reached from the north via the C-811 (Las Palmas to just east of Cruz de Tejeda), CV-15-6 and C-110, or from Tejeda via the CV-17-5. This road winds along the foot of the cliffs forming the north wall of the Barranco de Tejeda; it cuts down to the east through the lower part of the Roque Nublo Group as far as El Rincón, which is just 1 km horizontally – but 500 m vertically – from Degollada de Las Palomas.

Stop 5.25 Overview from Artenara   After parking in Artenara, walk past the church to the terrace of a café that overlooks Barranco de Tejeda and offers views of Roque Bentaiga, Mesa de Acusa and Mesa de Junquillo (the last two are in Fig. 5.12). All three are outliers of Pliocene and younger rocks, along with the long ridge from Lomo de los Marrubios (behind Roque Bentaiga as viewed from here) to Mesa de Junquillo. The much drier conditions, abundance of bare rock and sparser semi-arid vegetation of the south of the island, compared to the pine forests and subalpine meadows through which the road from Las Palmas passes, are very apparent from here. From the terrace, cliff exposures of Roque Nublo Group rocks, consisting of interbedded lavas and breccias, intruded by northwest-trending dykes, are visible to left and right. The slopes immediately below are terraced, but red (as a result of alteration) Miocene intrusive rocks emplaced in the centre of the caldera form the valley floor far below.

Of the various outliers, the slopes below Roque Bentiaga and the ridge behind, together with the lower part of Mesa de Acusa, are composed of well stratified sequences dominated by lava flows, with a well developed stepped or **trap** topography. The upper part of Roque Bentaiga and the crags to the south are composed of thick-bedded to massive breccias of the Roque Nublo ignimbrite type (see p. 124 and stop 5.26), but the upper part of Mesa de Junquillo is composed of an even more massive rock unit: this is an isolated remnant of the debris avalanche produced in the third collapse of the Roque Nublo volcano, which at one time must have filled the entire Barranco de Tejeda.

On leaving Artenara, take the second exit at the junction west of the

Figure 5.12 View southwest from Artenara into the lower reaches of Barranco de Tejeda. The canyon is 1 km or more deep in this area. The flat-topped plateaux within the canyon are Mesa de Acusa (to the right) and Mesa de Junquillo (middle distance, with north wall in shadow), erosional remnants of Pliocene rocks that once filled the canyon. Almost all other rocks in this view are part of the Miocene intrusive complex and caldera fill sequence; the margin of the caldera runs through the valley in the far distance beyond Mesa de Junquillo.

town (CV-3-1) and, after 3 km, turn left on the CV-3-4 for Mesa de Acusa.

Mesa de Acusa is another outlier of young rocks, forming an isolated platform perched high on the side of the Barranco de Tejeda after the most recent re-incision of the barranco. The top of the mesa is formed by a thick early Pleistocene lava flow similar to the flows of the Llano de la Pez formation at the head of Barranco de Tejeda (p. 120), but the most interesting rocks are exposed on the western and southern sides of the mesa, as the road (the CV-3-4) descends below the level of the top of the mesa. The road is very narrow and the most convenient way to view the following outcrops is to park in the village on the south side of the mesa and walk back along the road along its western side. **Caution: beware of traffic on this very narrow road.**

Stop 5.26 Western side of Mesa de Acusa   The lower part of the Mesa de Acusa sequence is composed of thick-bedded breccias that form well developed exposures above the road. It may be necessary to scramble up a few metres to examine these. These rocks are among the best examples of a very distinctive rock type, the Roque Nublo ignimbrites. These represent the opposite end of the ignimbrite spectrum from the very high-temperature ignimbrites of the Miocene sequence. They are characterized by abundant lithic and pumice fragments set in a weakly to moderately indurated tuffaceous sandy matrix; there is no evidence of compaction or welding. The lithic fragments are mainly basaltic-to-phonolitic lavas and scorias of the earlier Roque Nublo sequences. Among the most distinctive clasts are large irregular white-to-grey pumices with internal flow banding and ragged margins (Fig. 5.13). These have low vesicle contents and in some cases are better described as scoria clasts than pumices. They point to the volatiles that drove the explosive eruption of these rocks being derived externally rather than by exsolution as bubbles from the magma. Together with the low temperature of emplacement (from the lack of compaction and induration, and the fresh appearance of many of the lithic fragments) and the almost complete lack of Miocene rocks in the lithic clast assemblage, this implies that the volatiles were derived principally from groundwater heated by emplacement of magma at shallow depths. These rocks are therefore the product of hydrovolcanism.

Follow the CV-3-4 towards the reservoir (Embalse de Parralillo) in the barranco's floor. After about 3 km a wide parking place offers a spectacular view of the Mesa de Junquillo and the underlying Miocene rocks.

Stop 5.27 Embalse de Parralillo mirador   The Embalse de Parralillo mirador is built on a heavily jointed and altered crag of syenite. This is part of the large intrusive complex emplaced within the Miocene caldera after it

Figure 5.13 Large pale pumiceous clasts ("rags") within the Roque Nublo Group ignimbrites exposed in the roadcuts on the western side of Mesa de Acusa. The angular dark clasts are basaltic lithics, derived from the lower part of the Roque Nublo Group sequence.

formed and during emplacement of the dense swarms of inclined sheet intrusions that define the annular **cone sheet** swarm seen in the upper part of Barranco de Arguineguín and Barranco de Ayacata (pp. 109–110). The emplacement of these intrusions resulted in uplift of the caldera floor by at least 1 km; like many such calderas, it is therefore called "resurgent".

This mirador also offers perhaps the best roadside view of the thick sheet of debris-avalanche breccia that forms the 100 m or so of Mesa de Junquillo. This deposit has an irregular undulating basal contact, implying that it infilled a barranco system that, although not as deep as the present barranco, had been incised into the earlier Roque Nublo lavas below, after they had in turn infilled the earliest Barranco de Tejeda, formed during the extended period of erosion between the end of Miocene (shield-building stage) volcanism and the onset of growth of the Roque Nublo volcano.

From this mirador, continue on the road as it descends to the level of the reservoir and then down the deep canyon (up to 1300 m deep) that forms the lower reaches of the Barranco de Tejeda. The canyon cuts through the highlands formed by the massive intensely intruded and strongly altered ignimbrites and lavas of the caldera-fill sequence. The views are spectacular, but suitable stopping places are few, if you are travelling with two drivers and intend returning up the barranco (see below), swap drivers in San Nicolás de Tolentino; otherwise, change over part-way down so that both have the opportunity to take in the view.

The road finally emerges into the San Nicolás–Aldea basin, crossing the rim of the caldera as it does so; this is marked by a zone of intense greenish alteration. San Nicolás itself offers a variety of options for lunch. It is then possible either to drive south to Mogán and Maspalomas if time is pressing and you are based on the south coast, or north to Agaete and Las Palmas. However, the alternative of returning up the barranco is to be preferred, especially if you are in a four-wheel-drive vehicle (or are very confident on tracks), as the following alternative to returning to Artenara is then possible in good weather.

Follow the outward route in reverse as far as Embalse de Parralillo. At the upper end of the reservoir but while still on the south side of the barranco, look for a narrow track turning back and up hill to the right. This (the CV-17-9) leads via many very steep switchback turns to the col at the eastern end of Mesa de Junquillo (with spectacular views to both north and south) and the village of El Carrizal, at which point the road improves somewhat. From here onwards, the road climbs through a sequence of Pliocene (Roque Nublo Group) lavas and breccias, representing the early stages of growth of the Roque Nublo volcano, and overlying Roque Nublo ignimbrites. As discussed previously, this sequence infilled the ancestral Barranco de Tejeda, and the modern Barranco del Juncal, a tributary of the Barranco de Tejeda, is now cutting through the sequence again, resulting in the development of many spectacular cliffs.

The road eventually emerges (after 1–1.5 hours of strenuous driving) at a junction with the C-815 between Tejeda and Ayacata: from here it is possible to return either to Las Palmas or Tejeda by turning left or to the south coast by turning right through Ayacata and thence to San Bartolomé de Tirajana.

Northeast Gran Canaria
Post-Roque Nublo (late Pliocene to recent) basaltic volcanism in Gran Canaria

As noted in the introduction, post-Roque Nublo Group volcanism in Gran Canaria is concentrated in the northeast of the island and is dominated by alkali-basaltic and basanitic lava flows. The youngest of these flows are just prehistoric, and the area must be considered to be a probable site of future eruptions. Such eruptions, although they commonly cause extensive property damage, are normally not very dangerous to human life, because the lava flows travel slowly. However, there are thick sequences of sediments (notably the pre-Roque Nublo Group conglomerates seen in the roadcuts on the autopista from the airport to Las Palmas) beneath the surface in this part of the island and these are highly porous. The water table is on average about 200 m below the surface in this part of the island, and the sediments contain large amounts of water. As a result, intrusion of basic magmas at shallow depth, although not dangerous on its own, has sometimes produced violent steam explosions in the sediments. These produce deep craters rimmed by tuff rings of brecciated country rocks and basic lapilli and ash, known as maars. On Gran Canaria, future maar eruptions represent perhaps the most important short-term volcanic risk, because the northeast of the island is heavily populated.

This day's excursion will visit two of the best-preserved maar explosion craters in Gran Canaria, and also take in overviews from the highest point on the island, Pico de Las Nieves. It is relatively short and can therefore be combined with a walk in the Roque Nublo area or along the ridge between Cruz de Tejeda and Artenara, looking down on the recent volcanic activity in the north of the island and also taking in views down into the Barranco de Tejeda, or with a visit to the coastal outcrops west of Las Palmas described at the end of this section.

The Caldera de Bandama is reached by driving out of Las Palmas along the C-811 and turning left in Monte Coello on the CV-13-5 (the road system is being reconfigured at the time of writing and the route may change). However, the crater is a popular tourist excursion and is well signposted.

Stop 5.28 Summit of Pico de Bandama The Bandama complex consists of two components: a large scoria cone and the Caldera de Bandama itself. They are aligned in a north-northwest–south-southeast direction, which most probably reflects the orientation of an underlying dyke intrusion that fed both cones; there is no sign of a break between the two eruptions. An adjacent, slightly older scoria cone has yielded carbonized wood, which has given a ^{14}C age of 3500 years BP. Many other scoria cones of the northeast Gran Canaria volcanic field are visible from Pico de Bandama: Montañas Lentiscal, Tafira, Arucas and Jinámar, and the large coastal vent of La Isleta, which forms the island at the far end of Las Palmas. Other smaller vents and basanitic lava flows occur closer to Pico de Bandama.

Stop 5.29 Caldera de Bandama A short walk from the coach park at the northern side of the caldera leads down to some Guanche caves part way up the wall. It may be possible to examine the walls of the caldera close up. Otherwise, the deposits in the wall can be examined from the rim (binoculars may be useful).

The Caldera de Bandama is mainly excavated in bedrock and is about 1 km wide and over 200 m deep. The walls are mostly made up of Plio-Pleistocene lavas, although breccias (possibly lahar, debris-flow or debris-avalanche deposits) correlated with the Roque Nublo Group have been reported from near the very bottom of the wall. Around the rim, very fine-grain hydromagmatic ashes, with the buff to yellowish colour typical of the products of maar eruptions (as a result of alteration by the escaping steam), form the lower part of the eruptive sequence. Channels formed during the eruption (perhaps by water condensed from the eruption column) cut this sequence, and rare irregular "cauliflower bombs", produced by quenching of intruding magma by groundwater, are present. The hydromagmatic deposits are overlain by beds of black basaltic lapilli (well exposed along the road on the north side of the rim as well), indicating that, as is usually the case, the eruption began with steam explosions and then, as the supply of groundwater dried up, passed into a Strombolian type of activity. The floor of the crater is occupied by porous breccias and sediments that partially filled the explosion vent, which would originally have been much deeper still.

After rejoining the C-811, continue up to Cruz de Tejeda, where the CV-15-10 is followed across the Llanos de La Pez to Pico de Las Nieves.

Stop 5.30 Pico de Las Nieves[*] Pico de Las Nieves, the highest point on the island, along with the adjacent Pozo de Las Nieves, is composed of southeast-dipping breccias belonging to the Tirajana formation of the Roque Nublo Group. From this viewpoint, it is possible to see the arcuate arrangement of crags of Roque Nublo Group rocks around the plain of the Llanos de La Pez which is now occupied by the later lavas of the formation of that name; note in particular the high ridge of Roque Nublo Group rocks from Cruz de Tejeda to Artenara. This distribution of outcrops reflects the erosion out of the soft intensely altered core of the Roque Nublo volcano after the collapses, and the filling of the resulting basin by the Llanos de La Pez lavas and lacustrine sediments.

Stop 5.31 Roque Redondo More extensive views of Barranco de Tirajana and the mountains to the south, composed of Miocene rocks, can be obtained from Roque Redondo, a few hundred metres east of the radar station along a rather poorly maintained road. In particular, good views along the headwall of the Barranco de Tirajana, the kilometre-high cliffs of Agujerada, can be obtained; these expose a complete section through the Tirajana and Riscos de Chapín formations.

About 1 km north of the radar station is a junction: the excursion route turns east along the CV-18-3 to Cazadores and Telde. The route passes through the youngest rocks of the Roque Nublo Group, the Tenteniguada

---

* Coach-turning bay on west side of radar station; it may be necessary to park near the gates of the radar station and walk the last few hundred metres; if so, **beware of traffic.**

formation. This contains a few phonolitic lava flows and domes, but is mainly composed of phonolitic plugs that form prominent pinnacles, notably Roque de Tenteniguada itself.

About 6 km along the road from the junction, the road skirts around the Caldera de Los Marteles, another Holocene maar, which lies within a chain of scoria cones and smaller explosion vents aligned along a trend west-northwest to east-southeast; again, this probably reflects the orientation of the dyke or dykes that fed this group of eruptive vents.

Stop 5.32 East rim of Caldera de Los Marteles The deposits on the rim of the crater overlie an older scoria cone draped with slope talus. Unlike the sharp distinction between hydrovolcanic ash and later lapilli at Caldera de Bandama, the deposits associated with Marteles show a gradation from hydromagmatic ash breccia up through a transitional sequence of lapilli-rich beds, including composite (accretionary?) lapilli. This sequence provides good evidence for "drying out" of the eruption through time as the rate of groundwater supply decreased.

The return route to Las Palmas follows the CV-18-3 for 2 km east of Caldera de Los Marteles and then turns left down a switchback road into the Barranco de Guayadeque, a deep barranco incised into the Plio-Pleistocene lava sequence, thence to Ingenio and the autopista.

Pliocene and Pleistocene marine sediments and lava flows along the northeast coast As noted in the introduction to this chapter, northeast Gran Canaria has undergone slight uplift and, as a result of this, together with global fluctuations in sea level, there are exposures along the coast where pillow lavas and lava flows are interbedded with shallow marine sediments. The interaction of the lava with sea water and with the water-saturated sediments has produced interesting structures.

These stops can be added to the end of the previous itinerary. In this case take the autopista into the centre of Las Palmas and follow the signs for Guía and Gáldar (C-810) through the tunnel beneath the city centre. Alternatively, they can be visited at the start of the day or as part of a round-the-island tour if you start very early.

Stop 5.33 El Rincón From Las Palmas follow the C-810 and park in front of a high cliff just outside the town, at El Rincón. At the left-hand side of this cliff both shield-stage and post-erosional volcanic rocks are exposed in a 150 m-high palaeo-cliff. The first 50 m consists of strongly welded phonolitic ignimbrites of the shield stage of growth of the island (Miocene). About 40–50 m of conglomerates, sandstones, clays and volcanic ashes of the Las Palmas formation, representing the erosional gap between the two periods of volcanism, separate the shield-stage ignimbrites from Pliocene (early Roque Nublo) basaltic flows. The base of these flows shows pillow structures where they were emplaced into shallow water.

Stop 5.34 Barranco Cardones and Punta La Salina Follow C-810 to km-7.5 and turn towards the coast down Barranco Cardones to Punta La Salina. Marine erosion during the Quaternary is apparent at the end of Barranco Cardones (Fig. 5.14). Post-Miocene basalts forming a palaeo-cliff rest on a marine abrasion platform cut into Miocene phonolites. They are discordantly overlain by Quaternary tephritic lavas from the Montaña Cardones and Montaña Arucas volcanoes. The lavas represent two eruptive events separated by an interglacial sea-level maximum, corresponding to the oxygen isotope stage 11 (about 362 000–423 000 years ago), represented by marine sediments with marine fauna topped by a palaeosoil with continental fauna. The Montaña Cardones flow shows pillow structures at the bottom, evidence of abrupt cooling as they entered the sea.

The inferred succession of geological events corresponds to:

- The generation of marine abrasion platforms and cliffs on shield-stage formations during the Miocene.
- During a Quaternary interglacial period a post-erosional lava flows into the sea, forming a new coastal lava platform and a cliff.
- The last eruption occurred during a period of lower sea level and the lava flows on a palaeo-soil developed on the fossil beach, located 35 m above present sea level.

However, sea level during part of oxygen isotope stage 11 was as much as 20 m above present level and so the amount of uplift since emplacement of the Montaña Cardones flow may be as little as 15–20 m.

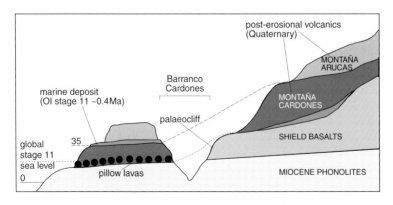

Figure 5.14 Sketch cross section showing field relationships between palaeo-cliffs, lava platforms and palaeo-soils in Barranco Cardones (stop 5.34). Sea levels are in metres relative to present sea level. (Modified after Meco et al. 2002.)

Stop 5.35 Punta del Camello: Montaña de Arucas lava flow   At km-10 of the C-810 road, the cliff is formed by a lava flow from the Montaña de Arucas volcano. The lava is a black phono-tephrite with prominent royal-blue crystals of hauyne, a light feldspathoid that floated in the lava and concentrated at the top of the flow. This lava, the Piedra de Arucas, has been widely used for monumental building (e.g. the cathedral of Arucas).

Turn inland to Arucas, one of the nicer towns in Gran Canaria (especially the cathedral and botanic gardens) and then return towards Las Palmas along the inland road through Tamaraceite.

Stop 5.36 Barranco de Tamaraceite   Past the town of Tamaraceite and in the main road to Las Palmas there is a road on the left to the arsenal of the naval base of Las Palmas. The road, which follows the final stretch of the Barranco de Tamaraceite at about 80 m above present sea level, cuts the Las Palmas formation, composed of alluvial phonolitic conglomerates and sandstones overlain with clays and volcanic ash, including at the top very shallow marine sediments representing the Pliocene marine transgression. The sedimentary formation is covered with a thick (40 m) basaltic flow of the Roque Nublo Group, separated in two distinct parts: the lower 25 m with well developed pillows and hyaloclastites, and the upper part of pahoehoe lavas. This is a spectacular example of a basaltic lava flowing into a shallow marine environment. Close inspection of the base of the lava flow in the roadcut reveals interesting textures and structures produced by chilling of the pillow lavas as they entered the water and also as a result of liquefaction of the underlying sediments (Fig. 5.15).

Figure 5.15   Roadcut in Barranco de Tamaraceite, exposing Pliocene marine carbonate sand disturbed by downward emplacement of pillows at base of overlying basaltic lava flow.

Chapter 6

Tenerife

Introduction

Tenerife is the largest of the Canary Islands, and has the third-highest oceanic-island volcano in the world after Mauna Loa and Mauna Kea in Hawaii (Fig. 6.1). It is also by far the largest of the highly active, shield-building-stage islands in the western half of the archipelago (see Ch. 1). The summit of Pico del Teide, 3718 m above sea level and thus the highest point in Spain, is over 1.5 km above the older subaerial volcanic rocks upon

Figure 6.1 Shaded-relief topographic image of Tenerife. Note the contrast between the deeply eroded Teno and Anaga Massifs (in the northwest and northeast respectively) and the younger parts of the island, dominated by the younger volcanoes and lateral collapse scars. (Image: GRAFCAN)

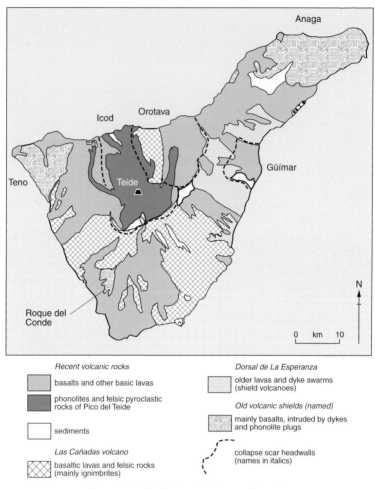

Figure 6.2 Simplified geological map of Tenerife.

which it is built and more than 7.5 km above the surrounding ocean floor. In addition to Teide, Tenerife has several collapse structures and by far the greatest volume of Quaternary felsic (trachytic to phonolitic) rocks of any of the islands.

The simplified geological map of Tenerife (Fig. 6.2) is based upon studies that showed that the island could be divided into two parts:

• The old volcanic edifices of Teno and Anaga, which form the eponymous peninsulas in the northwest and northeast of the island, and of Roque del Conde in the southwest. Coastal and shallow marine sediments found in boreholes around La Laguna, at the southwestern edge of Anaga, indicate that Anaga originally formed a separate island. These volcanoes were active between about 12 million and about 4 million years ago, and since that time have undergone deep erosion, producing spectacular terrain in which the internal structures of the volcanoes are well exposed.

- Since about 3 million years ago the central and southern parts of the island have been covered by younger rocks. This activity is centred upon the great summit depression, the Caldera de Las Cañadas (hereafter referred to as "the Caldera" to avoid the presumption that it is a caldera in the genetic sense). This has been characterized by alternating cycles of basaltic and felsic volcanism, whose products make up the Las Cañadas series, which forms the walls of the depression, and the younger formations that partly fill it and form the two volcanoes of Pico de Teide and Pico Viejo. The felsic rocks are mainly pyroclastics, and, as well as forming most of the rim of the Caldera, they form a sequence of ignimbrites and Plinian airfall pumice deposits in the Bandas del Sur to the south and in parts of the Tigaiga Massif to the north. Meanwhile, mainly basaltic volcanism has continued throughout on the lower slopes of the Las Cañadas volcano and to the east, forming the ridge of the Dorsal de La Esperanza.

It has long been assumed that activity has been continuous in Tenerife since emplacement of the oldest subaerial lavas in Roque del Conde, some 12 million years ago. This would imply that Tenerife is still in its shield stage of activity making the age sequence of islands from east to west irregular (see Fig. 1.2). However, recent work indicates that a break in volcanic activity on Tenerife may have occurred from about 4 million years ago. This implies that the Las Cañadas volcano is in fact a very large and particularly vigorous post-erosional volcano, comparable to the Pliocene Roque Nublo volcano on Gran Canaria, and also that the Dorsal de La Esperanza between Anaga and the Caldera is also a post-erosional feature.

Teno and Anaga, the peninsulas in the northwest and northeast of Tenerife respectively, are generally considered to represent separate, mainly basaltic, volcanoes. However, marked angular unconformities are also present within the sequences, and are to be seen at their most spectacular around Masca, in Teno. These unconformities indicate that the development of these parts of the island was much more complex than previously thought. They may well represent the scars of giant lateral collapses that occurred during the growth of the volcanoes.

Whatever the number of discrete periods of growth in Teno and Anaga, the main rock types present are sequences of lavas cut by many dykes and fewer intrusive plugs. In Teno these rocks are mainly of basaltic composition, but in Anaga, particularly at the top of the lava pile and in the core

139

of the intrusive complex along the north coast of the peninsula, a significant proportion of the rocks is of phonolitic composition. The spectacularly exposed dyke swarms (with up to 50% of outcrop formed by intrusions along the northern coast of Anaga) and the phonolitic plugs make an interesting counterpart to the basaltic fissures and phonolite lava domes of the Cumbre Vieja of La Palma. It is likely that Anaga, when active, was comparable to the Cumbre Vieja; thus, the intrusive rocks exposed there may give some insights into the deeper structure of the Cumbre Vieja. Since activity ceased, both peninsulas have undergone deep erosion, perhaps to depths of 2–3 km below the palaeo-surface. This has produced spectacular topography, particularly at the coasts, where seacliffs up to a kilometre high are found.

The centre of Tenerife is now mostly covered by phonolitic rocks of the Las Cañadas volcano. The many tunnels and boreholes excavated to drain groundwater for the island's water supply indicate, however, that the Las Cañadas edifice is underlain by a thick sequence of basaltic rocks. These are locally exposed around the periphery of this volcano, most notably at Roque del Conde on the southwest. Furthermore, the development of the Las Cañadas central volcano was accompanied by continued, mainly basaltic, volcanism in adjacent parts of the island, especially in the long ridge between the Caldera and Anaga: the Dorsal de La Esperanza. The Dorsal de La Esperanza is dominated by basaltic rocks. This fact, together with the occurrence of young (including eruptions in historical times) northeast-trending volcanic fissures and dykes along the crest of the ridge, led to the conclusion that it was a volcanic rift zone, perhaps comparable to the southern part of the Cumbre Vieja in La Palma. However, beneath these young scoria cones and lavas, much of this great ridge is formed by two large basaltic to phonolitic central volcanoes, upon which the more recent basaltic rift-zone fissure system has been superimposed. The older of these two, the Güímar volcano, forms the bulk of the older rocks on the eastern side of the Dorsal; dykes within this edifice radiate out from a centre at La Crucita, on the crest of the Dorsal at the head of the Güímar Valley. The younger volcano occupied the site of the present La Orotava Valley and its remnants are to be found around the rim of this valley. It was dominated by sequences of basaltic lavas, with some phonolitic lavas and pyroclastic rocks.

On the flanks of the Dorsal, to north and south, are two broad steep-sided valleys, La Orotava to the north and Güímar to the south (see Fig. 6.2). Another similar valley, the Icod Valley, lies to the north of Teide itself. Although these valleys have been greatly modified by erosion and partial re-filling by younger sequences of rocks, an origin in giant lateral collapses is now well established, particularly for the La Orotava and Icod valleys, since debris-avalanche deposits have been imaged (using deep-sea side-scan sonars) off shore from both (see Fig. 1.4). Little of the La Orotava volcano now remains (which is partly why it has not been recognized until very recently) as a result of the lateral collapse that produced the La Orotava Valley and the debris-avalanche deposits off shore.

The phonolitic volcanic rocks in central Tenerife are distributed around the Caldera, the large (16×9 km) depression in the centre of the island. Much of this sequence is phonolitic in composition and was erupted in violent explosive eruptions that deposited thick units of ignimbrite. The **distal** parts of the ignimbrite sheet, and associated lahar deposits, Plinian airfall deposits and other rocks form the plain of the Bandas del Sur in the south of the island. Correlation of these rocks with the **proximal** rocks of the Las Cañadas edifice in the interior of the island has proved problematic, partly because of deep erosion in the intervening slopes, which has removed much of the upper part of the sequence, but also because this region has not been mapped in detail. Studies of the proximal deposits have focused upon the sequences in the caldera wall.

The Las Cañadas sequence in the walls of the Caldera is divided into the Lower Group of basaltic and minor phonolitic lavas, scoriaceous units and breccias, and the Upper Group divided into three formations: Ucanca, Guajara and Diego Hernández. These three units are discussed in more detail below (pp. 153–154).

The origin of the Caldera de las Cañadas itself is one of the great controversies of the geology of Tenerife. For this reason the term "caldera" as applied to this depression should at present be used only in its original Spanish topographic sense (a large depression at the top of, or on the flank of, a mountain). One school of thought holds that the Caldera was produced by coalescence of the three overlapping calderas produced by the large-volume eruptions that produced the pyroclastic rocks of the Ucanca, Guajara and Diego Hernández formations. The other school of thought points to the absence of a northern wall to the Caldera, apart from the short Fortaleza section in the northeast. The missing section of the north wall is at the head of the Icod Valley, which, from the offshore evidence in particular, appears to be a lateral collapse scar (see Fig. 1.4). Furthermore, limited evidence from boreholes and water tunnels indicates that the floor of the Las Cañadas depression, beneath the later Teide–Pico Viejo complex, slopes steeply northwards and is covered by coarse sedimentary or debris-avalanche deposit breccias. This evidence indicates that the Caldera is either part of the Icod collapse scar or that it was produced by erosion that enlarged the collapse scar, as occurred at Taburiente in La Palma.

Whatever the mechanism and timing of its formation, after the Las Cañadas depression developed, it has been mainly filled by the later rocks of the Teide–Pico Viejo complex. These have also filled most of the Icod Valley to the north (see Fig. 6.2). The lower parts of the sequence filling the Caldera are poorly exposed but have been studied in boreholes and the few water tunnels that penetrate the walls of the Caldera. It appears to be composed of basalts and alkali basalts at the base of the sequence, without well defined volcanic cones; these indicate that there was no major shallow magma chamber at this time and that the rocks were erupted from many small vents. Later the two cones of Teide and Pico Viejo began to develop and erupt more evolved magma compositions.

The two peaks of Teide and Pico Viejo are now the highest points in the Canary Islands. In the most recent stage of their growth, the summit vents of both volcanoes are characterized by phonolitic to intermediate (tephritic) magmatism. Many vents on the flanks have also erupted pho-nolitic lavas and pyroclastic rocks, defining a shadow zone within which no basic magmas have been erupted, although basic magmas have been erupted from many vents outside this zone on the floor of the Caldera. This has been interpreted as indicating the presence of a stratified magma chamber beneath the shadow-zone area, containing tephritic magmas beneath a top layer of phonolitic magma, which traps the denser basaltic magmas and does not allow them to erupt in this region. Most eruptions of Teide and Pico Viejo are small and dominated by lava flows. However, at least one substantial explosive (sub-Plinian) eruption has occurred in the recent past: the Montaña Blanca eruption, dated at about 70 BC. Depos-its from this eruption are well exposed and easily accessible, as discussed below. Other phonolitic pumice layers, most probably also derived from the Teide–Pico Viejo complex, occur interbedded with the young basaltic scoria and lapilli beds in the Dorsal de La Esperanza. These deposits indi-cate that the Montaña Blanca eruption is only the youngest of several such Plinian or sub-Plinian explosive eruptions.

Today, the Teide–Pico Viejo complex represents perhaps the most important direct or magmatic volcanic hazard on the island, because of these relatively infrequent but potentially damaging explosive eruptions. Furthermore, the northern side of Teide in particular is extremely steep and could undergo lateral collapse. Such a collapse might be relatively small, involving only the steep upper cone, but nonetheless could threaten the whole of the Icod Valley, or it could conceivably be a giant lateral col-lapse affecting the whole region between Teno and the La Orotava Valley. However, no clear evidence has yet been found of precursory deformation similar to that found in the Cumbre Vieja of La Palma, and so the latter must be considered to be more likely to fail in the geologically near future.

Logistics on Tenerife

Tenerife has relatively good bus services, but the distances between the main towns mean that taxis are a very expensive way to get around. Hire cars, in contrast, are plentiful and very cheap. Thus, although the consequence is that walks generally have to be circuits or involve retracing your steps, the balance of advantage lies with using a hire car for most if not all of the time in Tenerife.

Tenerife is sufficiently large that, even with the widespread availability of hire cars, there is an argument to be made for using two bases while visiting the island, one in the northeast and one in the southwest (Fig. 6.3). Simply to drive the main circuit around the island, taking both the northern and southern autopistas and the C-820 around the western side, takes most of a day. For those who want to see more of the local life, La Laguna and Vilaflor are the best centres to use if following the two-base approach. If you prefer to stay in one of the main tourist centres, Puerto de la Cruz on the north coast is much more pleasant than any of the south-coast resorts, especially if you spend your evenings in the older western end of the town. It is also relatively close to most of the main areas of geological interest in Anaga, Teno, Valle de La Orotava and the Caldera. However, it will take you at least two hours' continuous driving to travel from Puerto de la Cruz to the Bandas del Sur or the Aeropuerto Reina Sofia, the most likely point of arrival and departure.

The Parador in the Caldera is of course the most central location of all, if you can afford it, but it is at over 2000 m elevation and many people find it difficult to sleep at that altitude. Altitude is also an important consideration when in the field, especially on the crest of the Dorsal de La Esperanza and in the Caldera. The latter is for the most part close to being a true desert, and both sunburn and sunstroke have to be carefully guarded against during the day; and it can quickly get very cold at night or in bad weather. Roads above about 2000 m can be blocked by snow in winter. The summit of Pico del Teide is high enough that altitude sickness is common among visitors; do not exert yourself up there and especially do not take the path up (use the cable car) unless you are very fit and used to mountain trekking. In contrast, clouds often descend on the northern slopes of the island, especially in Anaga, so be prepared for cold and wet conditions and careful navigation there.

143

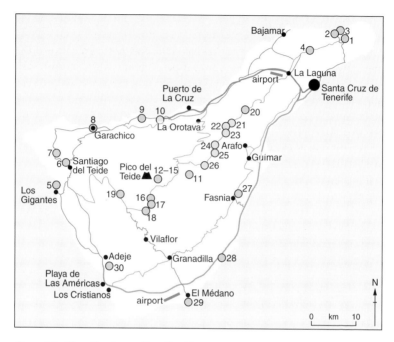

Figure 6.3 Map of locations on Tenerife, with major towns also indicated. Only selected roads are shown in the north and east of the island; autopistas indicated by heavier lines.

Anaga

As discussed in the introduction, Anaga is (along with the Teno and Roque del Conde), one of the three old parts of the island and has been very deeply eroded, to produce a wild landscape of high cliffs and deep valleys falling away to the sea. The northern, windward side of Anaga is the wettest part of the island, with a layer of orographic cloud often shrouding the peaks. It is heavily vegetated; with tree heathers and laurel forest are still extensively developed. Many of the best and most easily accessible exposures are in fact in roadcuts. In consequence, although there are many paths and excellent walks in good weather, under most conditions the best way to see the rocks of Anaga is by following the road circuit described here. Anaga is rarely visited by most tourists, but is popular with the locals; hence, it is best to visit it on a weekday and should definitely be avoided on fiestas. The car tour described here takes half to two thirds of a day; it can be combined with visits to Santa Cruz or (better) La Laguna, or a drive along the Dorsal de La Esperanza route described below.

The main structural trend in Anaga is east–west, parallel to but not co-incident with the high ridge of the peninsula. The southern side is formed primarily by sequences of southward-dipping lavas, whereas the north slope is mainly formed by dykes and small but prominent intrusions, cutting altered lavas and other volcanic rocks. This area therefore exposes the deeply eroded core of the Anaga volcano or volcanoes, the summits of which would have been well to the north of the present high ground. The original northern slopes are missing, removed by erosion of the windward coast and possibly also by repeated lateral collapses to the north.

Beginning in Santa Cruz de Tenerife, which is built upon relatively recent lavas erupted from the Dorsal de La Esperanza, drive along the Avenieda Marítima towards San Andrés and the Playa de Las Teresitas (an artificial beach, popular with the locals, made from sand imported from what was at the time Spanish Sahara on the African mainland). At the eastern end of Santa Cruz harbour, the road crosses onto the deeply eroded terrain of the old Anaga volcano. The cliffs above the road are now composed of thick sequences of southward-dipping basalt lavas. Continue along the south coast of Anaga to San Andrés, and then up the steep and winding road (TF-112) that leads to the crest of Anaga.

This southern side of Anaga is dominated by basalt lava and scoria sequences, but, as the road climbs, these are cut by more and more basaltic dykes and phonolite plugs, forming pale and heavily jointed pinnacles that the road loops around crazily. If you decide to stop to have a closer look at any of these, be very careful of the traffic. As you climb, there is a marked change in the vegetation, from mainly cultivated Mediterranean scrubland to laurels and tree heathers, reflecting the increase in rainfall.

Stop 6.1 El Bailadero Just before the tunnel through the summit ridge is reached, a road junction offers two choices. In clear weather, turn off to the left (TF-1122), initially on the road to Tegueste but then taking another turn back to the east onto the narrow road (TF-1123) to the summit ridge and the prominent restaurant (El Bailadero) and mirador on the ridge. This offers spectacular views to the north, looking down precipitous slopes to the north coast of Anaga. From this distance, the most prominent features of the slopes are the many pale pinnacles formed by phonolite intrusions that cut the poorly exposed lavas and dykes. Continuing east along this road leads to further miradors, also with spectacular views to north and south, and the top ends of the paths down the slopes of Anaga (see below). Roadcuts along the way expose phonolitic lavas and scoriaceous pyroclastic rocks. These form the topmost volcanic unit of the Anaga volcano, a thin cap on top of the older basaltic sequence. It is likely that these rocks are the remnant of a thicker and more extensive sequence fed by the many phonolitic intrusions exposed to the north.

Returning to the road tunnel (or continuing through it, in cloudy conditions), the descent to Taganana along the TF-1124 follows a winding route down the very steep slopes. Some small miradors offer the only opportunities to stop and examine the rocks, including some of the prominent phonolite pinnacles. As one proceeds down hill and to the north, there is a marked increase in the number of plugs and also in the number of mostly basaltic and mainly east–west trending dykes. Very few of these cut the phonolites, supporting the inference that the phonolite intrusions were emplaced late in the history of the volcano and most probably contemporaneously with the phonolitic volcanic rocks that cap Anaga.

Stop 6.2 Taganana village   By the time that Taganana village is reached, the roadcuts are dominated by moderately altered, brownish to greenish intrusive rocks. A roadcut at the entrance to Taganana village, opposite a long layby, is formed almost entirely of basaltic to phonolitic dykes. These formed part of the east–west dyke swarm that lay at the heart of the Anaga volcano. This exposure offers an interesting insight into the likely structure and state of alteration of the interior of comparable younger volcanoes, such as parts of the Dorsal de La Esperanza on Tenerife and also the Cumbre Vieja on La Palma, at depths of 1–2 km below their summits. The alteration mineral assemblages are distinctly unimpressive, with much of the rocks being partly altered to clay minerals that have in turn oxidized on exposure (hence the widespread rusty colours) and the only obvious new minerals being white carbonates and zeolites in veins. The alteration reflects the circulation of moderately heated groundwater through the volcano; it occurred at markedly lower temperature than the metamorphism seen in the dyke swarms on Fuerteventura, reflecting the fact that the latter were uplifted and exposed by much deeper erosion than is encountered in Anaga. On the other hand, the development of clays means that this lower-temperature alteration weakens the rocks much more than the higher-grade metamorphism, and may contribute significantly to the instability and collapse of such volcanoes.

Stop 6.3 Roque de las Bodegas   The road and the Anaga dyke swarm continue to the coast at Roque de las Bodegas. Here the older rocks are overlain by coarse palaeo-screes that are presently being undercut by the sea. The screes contain abundant phonolite fragments, often with plenty of pyroxenite and amphibolite xenoliths, indicating that the cap of phonolitic rocks on top of Anaga may once have been much more extensive. Roque de las Bodegas features fish restaurants and is a good place for lunch.

Stop 6.4 Pico del Inglés: overview of La Laguna   Return along the TF-1124 to Taganana and the summit ridge, then turn east along the ridge road (TF-1123) towards La Laguna. There are many vantage points along the route. Of these, perhaps the most worthwhile is that at Pico del Inglés. This provides an excellent view over the Vega de La Laguna. This remarkably flat plain was formed when a deep valley draining the southwestern side of Anaga was blocked by the growth of the Dorsal de La Esperanza and subsequently filled by volcanic ash and lavas, and by sediment eroded from Anaga. At the time of the Spanish conquest, the Campo was occupied by a marshy lake (hence the name of the town). However, this was subsequently drained and the lake bed is now one of the most fertile agricultural areas on the island.

The road continues down the southern slope of Anaga, through the main remaining laurel forest on the island, across La Vega de La Laguna; if heading for the autopista, you may wish to turn left in Las Mercedes along the eastern side of La Vega de La Laguna and bypass the town, unless you are attracted by its many examples of Spanish colonial architecture.

Teno and Garachico
The Tigaiga Massif and the lower La Orotava Valley

The Teno Massif at the northwestern corner of Tenerife is by no means as large as Anaga, although it is characterized by equally or even more extreme terrain. Unlike Anaga, younger volcanic rocks, including some historical-period lava flows, have been erupted in the area, filling valleys and pouring over coastal cliffs. Thus, in one day a drive through Teno can be combined with a visit to the town of Garachico on the north coast, and stops along the coast to the east, ending in the lower La Orotava Valley at Puerto de la Cruz. During the tour described here, some of the oldest rocks in the island are to be seen, as well as many younger ones, including the products of the most recent eruption in Tenerife.

The itinerary described here begins in Los Gigantes ("the giants"), the most upmarket of all the resorts along the west and south coasts of Tenerife. It can be reached from the south via the TF-6237 (the coast road) or, higher up the western side of the island, the TF-822, TF-820 and TF-6281, or from Puerto de la Cruz via the Autopista del Norte and the TF-820. The part of the itinerary through Teno coincides with the highest humidity and is often cloudy and damp, so pick a good day for this section. Other parts, especially the traverse along the north coast, are often much clearer.

Stop 6.5 Los Gigantes Los Gigantes lies at the northern end of that part of the island (see Fig. 6.2) covered by the lavas and pyroclastic rocks of the Las Cañadas volcano. These relatively recent rocks have not been deeply eroded, so that high coastal cliffs have not developed. It is therefore easy to reach the coast, and from here or from one of the many terraces in Los Gigantes it is possible to look north to the cliffs that give the town its name (for a closer look, take a boat excursion from Los Gigantes; Fig. 6.4). These seacliffs, near vertical and up to more than 500 m high, are among the highest in the Canaries. They expose shallowly southwest-dipping basaltic lavas of the Teno Massif, cut by many prominent southwest-trending dykes. Between 6 million and 3 million years old, these lavas represent a section through the southern side of the Teno volcano, whose core lies farther to the northeast.

Leaving Los Gigantes, take the TF-6281 towards Tamaimo. The road climbs the Barranco de Tamaimo on the southern edge of the Teno Massif. The walls of this barranco are formed by deeply eroded old lavas of the Las Cañadas edifice to the southeast and by rocks of the Teno Massif to the northwest, but the barranco floor is formed by much younger lavas that have partly filled it. Although lavas will doubtless flow this way again, the village is on the only reasonably flat ground available. Typical rocks of the Teno Massif are exposed in the crags at the head of the Barranco de Tamaimo: basaltic lavas cut by a few dykes and, to the northwest, a pale yellow phonolite plug. However, phonolitic rocks are a much less significant component of Teno than either Anaga or the Las Cañadas volcano.

At the head of the Barranco de Tamaimo, the road enters a small basin, in which the town of Santiago del Teide is situated. This basin formed in

Figure 6.4 The cliffs of Los Gigantes, viewed from the south. Note many dyke intrusions extending up the cliffs.

the same way as the much larger La Vega de La Laguna basin in which La Laguna is situated: the growing Las Cañadas volcano blocked a valley incised into the Teno Massif, which then filled with sediments and ponded lavas. The good (and, perhaps more importantly, well watered) agricultural land thus formed has led to the growth of the settlement, despite the fact that it lies directly on the axis of the northwest rift zone of the volcano. In the town the route turns left onto the TF-1427, which rapidly climbs up into the Teno Massif to the west.

Stop 6.6 Mirador de Arasa   About 2 km from Santiago del Teide, the road reaches a col (with a car-park) from which there are views east and west. To the east, the slopes leading up to the Caldera and Teide are covered by the very active northwest rift zone. The terrain is a continuous mass of basaltic scoria cones and lava flows, some very fresh despite the humid conditions. The most recent eruption on Tenerife, the Chinyero eruption of 1909, formed lava flows that are visible southeast of Santiago del Teide.

The spectacular view to the west from this col, towards the village of Masca, is of the deeply incised Teno Massif. The near-vertical barranco walls, several hundred metres high, expose basalt lavas (including many rubbly flows) cut by many dykes of olivine basaltic through basaltic to trachybasaltic and (rare) phonolitic compositions. At least one important angular unconformity, which also truncates many of the dykes, is clearly visible; it runs through the village of Masca and dips steeply northwards, separating older south-dipping basalts to the south from younger flat-lying rocks to the north. It is important because it indicates that the Teno Massif contains at least two discrete volcanic edifices, separated by a marked period of erosion (or possibly a large-scale lateral collapse) that incised a kilometre or more into the older volcano.

The route continues down the very convoluted road to Masca. Many dykes are visible in the roadcuts and barranco walls, but, if you do decide to stop for a closer look, take great care and do not block the passing places. The best views down into the very deep Barranco de Masca are at larger parking places in the village itself. The unconformity noted above is crossed at the Mirador de la Cruz de Hilda, just beyond Masca on the road to Buenavista del Norte (TF-1427).

Stop 6.7 Carrizal Bajo overview   About 3 km beyond this mirador, another much smaller mirador offers a splendid view west and north over the valley containing the settlement of Carrizal Bajo. The younger sequence of rocks seen near Masca is cut by an unconformity that represents an ancient barranco, with steep walls clearly visible beyond Carrizal Bajo, filled by thin lava flows and beds of basaltic pyroclastic rocks. Again, two groups of dykes are visible, the older group being cut by this unconformity. It follows that the Teno Massif contains at least three discrete sequences of volcanic rocks, and in fact has a very complex growth history.

A short way beyond this mirador, the road crosses the col at the top of Tenerife. For much of the time this is marked by a transition into the cloud that hangs over northern Tenerife in most weather conditions; but even if the weather is clear the change in microclimate will be evident from the sudden increase in the density of vegetation. Beyond the col, the road descends a barranco to the Barrio de Palmar. The walls of this barranco are formed by Teno rocks, with prominent dykes exposed north of Barrio de Palmar. However, the barranco floor is occupied by a northwest-trending group of young scoria cones and lava flows. These represent the continuation of the northwest rift zone through Teno.

These younger rocks continue to the coast where the route turns east in Buenavista del Norte, onto the TF-142 to Garachico. The coastal platform is formed by the younger lavas, which originated partly from vents on the platform (Montaña Taco, north of the road, is a young scoria cone used as a reservoir) and partly from vents at the top of the old and now isolated seacliffs eroded into the Teno Massif. Montaña Taco, north of the road, is a young scoria cone serving as a reservoir. The fringe of recent lavas has made this coastline accessible and inhabitable, but further eruptions may cause much damage, as occurred at Garachico in 1706 (stop 6.8).

Stop 6.8 Garachico Garachico itself is on a very narrow part of the coastal platform, just below the palaeo-seacliffs cut into Teno rocks. In January 1706, lava flows from an eruption on the northwest rift zone descended the cliffs and destroyed much of the town. Even more damagingly, the flows filled the harbour, which had been one of the best in the islands, and destroyed the town's raison d'être. Although the town was rebuilt after the eruption, trade moved to Puerto de la Cruz and Santa Cruz de Tenerife. Consequently, Garachico became fossilized and today it has some of the best examples of eighteenth-century Spanish streets anywhere. The oldest building in the town is the Castillo de Fortaleza de San Miguel, an early sixteenth-century fortification on the eastern side of the harbour. This was built to protect the harbour, and the headland of older lava on which it was situated was one of the few parts of the town to escape the eruption of 1706.

East of Garachico, the TF-820 continues along the foot of the Teno Massif palaeo-cliffs (partly covered by younger lavas) on the coastal platform formed by these younger lavas. However, 6km to the east, shortly before the town of Icod is reached, the road enters the Icod Valley, and the older rocks disappear at a marked escarpment, running inland, which forms the western wall of the valley. The Icod Valley is an old lateral-collapse structure filled with younger lavas and soft sedimentary (and possible debris-avalanche deposit) breccias. These breccias, known as the Mortalon by the excavators of the water galleries, are very soft and are a likely failure surface in future collapses. The presence of these breccias is alarming considering the spectacular 20° slope up to the Teide–Pico Viejo complex at the head of the valley.

Stop 6.9 Playa del Socorro roadcut   About 8 km east of Icod, the TF-820 crosses out of the Icod Valley and passes along a narrow coastal platform at the foot of the old Tigaiga Massif (see Fig. 6.2). For a distance of 3 km in the western part of the Tigaiga Massif, the road passes along the foot of a massive bed (up to 40 m thick) of beige, partly welded, phonolitic ignimbrite. This is the largest ignimbrite in northern Tenerife, and has an Ar/Ar radiometric date of about 1.2 million years. Beneath this, exposed in a roadcut 5 km after the turn to San Juan de la Rambla, 3 km after the road crosses Barranco de Ruiz, and just before the turn to Playa de Socorro, is a spectacular rock unit well worth stopping for. However, be careful: this stop is on a busy road, and you will have to stop in a long but narrow parking bay, so beware of traffic.

At this point, the roadcut is excavated in a thick debris-avalanche deposit. It contains sheared blocks of lava and pumiceous phonolite breccia, up to 10 m across, in a massive, relatively coarse-grain matrix (N.B. the matrix is not muddy, so the deposit is not a lahar). The deposit is typical of the distal facies of debris avalanches of relatively moderate size, generated by lateral collapses of about the same size as the 1980 collapse of Mount Saint Helens. The likely source is an early volcano within the Las Cañadas edifice, perhaps that represented by the Ucanca formation exposed in the walls of the Caldera (see stop 6.18, p. 163).

Continuing eastwards along the TF-820, further exposures of pumiceous rocks occur at the base of the western wall of the La Orotava Valley, perhaps the largest (and certainly the most spectacular) lateral-collapse structure in Tenerife. From the road, the walls of the collapse structure are clearly visible to the south (west wall, leading up to the Caldera and the post-collapse edifice of Teide) and east (deeply eroded east wall, cutting rocks of the Güímar and La Orotava edifices). The floor of the valley is now mostly covered by a post-collapse sequence of lavas and pyroclastic rocks, partly derived from the Teide–Pico Viejo complex but mainly from the youngest vents of the Dorsal de La Esperanza. These are mainly alkali basaltic lavas, scoria and lapilli; lapilli are especially common in the upper valley, closer to the volcanic rift zone along the ridge to the south.

Stop 6.10 Punta El Guindaste coastal section Beneath these young lavas, and exposed along the coast between Punta El Guindaste and the western end of Puerto de la Cruz (access via the TF-1823 or the coast road through Puerto de la Cruz), is a sequence of bedded conglomerates, coarse sandstones and muddy debris-flow or mudflow deposits. These are among the few exposures of the Mortalon, although they are perhaps atypical in being relatively rich in boulder- to gravel-size sediment and relatively poor in mud, suggesting deposition in an alluvial fan. Nevertheless, given the easy erosion of these deposits by the sea, it is sobering to think that the whole of Teide is most probably built upon similarly weak foundations.

The Cañadas volcanic edifice and Teide
The Caldera de las Cañadas

A visit to this central and most spectacular area begins by climbing to the Caldera on winding roads from the south (C-821 via Vilaflor and the Boca de Tauce), the north (C-821 from La Orotava) or along the Dorsal de La Esperanza (C-826 to La Laguna). The geology of the Dorsal is worth a visit in its own right and is described separately below. Since conditions for photography in the Caldera and especially of Teide itself are in general best if the visit begins in the east in the morning and ends in the west in the evening, the route described here begins at La Orotava and ends at the Boca de Tauce in the southwest. Also, if you want to go up Teide itself via the cable car, it is best to head directly to the cable-car station first thing in the morning and return to El Portillo to continue this itinerary afterwards.

The Caldera is entirely within the Parque Nacional del Teide and there are therefore firm travel restrictions off the major roads. No collection of rocks or plants is allowed. Paths accessible to the public are generally well marked and the main routes are described in Rochford's walking guides. The main itinerary described here follows the main road with some short walks and one longer, steeper walk off it; notes on some longer walks follow. The cautions in Chapter 2 about carrying water and plenty of sun protection apply with special force to the Caldera because it is so high and, for most of the year, so dry. It can also be very cold, especially at night.

The route to the Caldera from Puerto de la Cruz follows the C-821 south through the town of La Orotava. In clear weather, the road up the slope of the valley affords excellent views of the cliffs to the east and south. These cliffs are mainly composed of brown- to reddish-weathering basalt lavas, scoria and epiclastic breccias, but at the top of the cliffs are pale-weathering trachybasaltic to phonolitic lavas. Both of these units belong to large volcanoes in the Dorsal de La Esperanza sequence, discussed below.

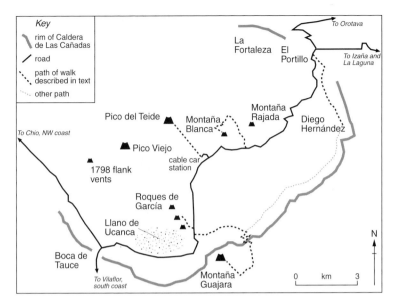

Figure 6.5 Routes and points of interest in Caldera de las Cañadas.

Younger scoria and lapilli units of the rift zone that runs along the crest of the Dorsal de La Esperanza are draped over the top of them in places.

The C-821 exits the La Orotava Valley at its southeast corner and enters the Caldera at the junction with the C-824 and the Parque Nacional de Teide visitor centre (at El Portillo; see Fig 6.5 for the route and various paths). The main rock units can be seen from the road as follows:

- The older rocks of the Las Cañadas volcano form the walls of the Las Cañadas depression, in the distance to the east and south of the road. These walls are very steep and it is clear that the depression was originally much deeper, and has filled up only relatively recently as the exits from the depression have been blocked by the growth of the Teide–Pico Viejo complex. The sequence in the walls is summarized in Figure 6.6 and Table 6.1. The Upper Group of rocks varies laterally along the wall, with the oldest (Ucanca formation) rocks mainly in the west and the youngest (Diego Hernández formation) rocks entirely confined to the eastern wall of the Caldera. The eponymous cañadas are the small sedimentary basins, filled with sand and gravel, that have formed around the depression behind the barriers formed by successive lava flows. The largest is the Llano de Ucanca, just west of the Roques de García. These are the line of oddly shaped pinnacles, formed by a spur of rocks extending north from the southern wall of the depression, that divide the floor of the Caldera into eastern and western halves.

153

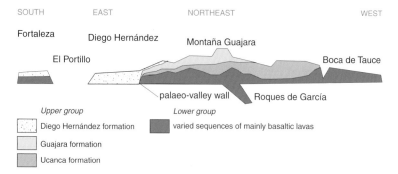

SOUTH   EAST   NORTHEAST   WEST

Fortaleza

Diego Hernández

Montaña Guajara

El Portillo

Boca de Tauce

palaeo-valley wall   Roques de García

Upper group   *Lower group*

Diego Hernández formation   varied sequences of mainly basaltic lavas

Guajara formation

Ucanca formation

Figure 6.6 Structure and stratigraphy (not to scale) of the wall of Caldera de las Cañadas, as viewed from the Pico del Teide (see Table 6.1). Many dykes and inclined sheet intrusions cut the older rocks, especially the Lower Group and the Ucanca formation.

Table 6.1 Summary of stratigraphic units exposed in the cliff walls of the Caldera de las Cañadas (see also Fig. 6.6).

Group	Formations	Lithologies
Upper Group	Diego Hernández	Non-welded pumiceous ignimbrites with interbedded ashfall pumice deposits and lava flows, intercalated with scoria cones
	Guajara	Thick phonolitic pyroclastic units, often strongly welded to lava like. Interbedded with non-welded ignimbrites, ashfall pumice beds and breccias. Many units are lens shaped because of deposition in palaeo-valleys
	Ucanca	Phonolite lavas and pyroclastic units: latter usually unwelded pyroclastic rocks (ashflows and airfall deposits). Rare basaltic lavas at top.
Lower Group		Varied sequences of mainly basaltic lavas, breccias and sediments, with many unconformities separating different units; probably represents several volcanoes.

- Teide and Pico Viejo are the two volcanoes that presently form the northern boundary of the Caldera depression, although they originally grew within it. Teide, the eastern volcano, is much the larger, and Pico Viejo is only visible west of Roques de García.
- The very rough floor of the Caldera is mostly covered by lava flows that originated from Teide and from vents on its flanks such as Montaña Rajada and Montaña Blanca (the road skirts around the southern side of both of these, passing through spectacular obsidian block flows). However, a few small lava flows originated from the many small scoria cones on the floor of Las Cañadas.

Stop 6.11 The cliffs of Diego Hernández   About 1.5 km south of the visitor centre a car-park on the right marks the start of a footpath that leads south over fairly flat terrain along the foot of the eastern wall of the Caldera. The first part of the walk leads through lava flows and past the lavas and scoria cones that drape the eastern wall. This area is the westernmost end of the

154

volcanic rift zone that runs east-northeast along the crest of the Dorsal de La Esperanza. In places, erosion has exposed the dykes that fed these vents. These dykes are resistant to erosion and they appear as near-vertical walls standing proud of the softer rocks through which they were intruded. Farther south, the cliffs are composed of a remarkably well bedded and flat-lying sequence of white to pale yellow pyroclastic rocks, also cut by darker, harder dykes and irregular **sills**. The pyroclastic rocks are airfall pumice beds and unwelded ignimbrites of the Diego Hernández formation, the youngest of the three main units that make up the upper walls of the Caldera (Fig. 6.7). After about a 3 km walk, the cliff curves around to the west and a prominent spur (Risco Verde) marks the western boundary of the Diego Hernández formation. The boundary is actually a steeply eastward-dipping contact against thick, faintly greenish phonolitic lavas and intensely welded ignimbrites of the Guajara formation, the second formation of the Upper Group (see Table 6.1). Although the contact looks superficially like a fault, it is in fact the steep wall of a deep palaeovalley that has been filled in by the rocks of the Diego Hernández formation: the ignimbrite flows ponded in this valley to produce the characteristic flat-lying bedding within the younger formation.

Continue around the Caldera's southern wall to the Parador Nacional, or return to the car-park and continue along the road, which skirts thick jagged phonolite lava flows erupted from the eastern side of Teide, part of the sequence that has filled the Caldera. These flows are young (15 000 to only a few thousand years old) and well preserved. Between the flows are patches of cream to pale greenish pumice, most belonging to the 2000-year-old Montaña Blanca eruption.

Figure 6.7 Pyroclastic rocks of the Diego Hernández formation exposed in the eastern wall of Caldera de Las Cañadas, cut by basaltic dykes and sills that feed scoria cones at the clifftop.

Montaña Blanca

Several paths leave the road on the south side of Montaña Rajada and Montaña Blanca, the most useful being that which forms the lower part of the steep path up the eastern side of Teide, the start of which is marked by a notice on the right-hand side of the road, south of Montaña Blanca itself. There is limited parking space on the left (south) side of the road, but on holidays in particular you may have to turn back and park a little farther east, in a larger parking area surrounded by a landscape draped by pale greenish to white pumice from the Montaña Blanca pumice.

Montaña Blanca is the site of the most recent major explosive eruption on Tenerife, which occurred about 70 BC and would almost certainly have been witnessed by Guanches. Because of the arid climate at this altitude and the lack of erosion within the Caldera, the products of the eruption have been exceptionally well preserved and are one of the finest examples of a small-volume sub-Plinian eruptive sequence anywhere. Deposits from the eruption cover much of the northeastern part of the floor of Las Cañadas and have been identified as far away as the central La Orotava Valley. The walk up to the summit of Montaña Blanca at 2600 m elevation takes about four hours in total; there is no shade whatsoever on the route, so be especially careful to carry sufficient water and sun protection.

The path ascends from the road between El Culatón, an obsidianic lava flow with steep sides and a ridged top surface (a **coulée**), which was emplaced late in the Montaña Blanca eruption, and the earlier El Tabonal flow to the east. El Tabonal, the first component of the Montaña Blanca eruption, overlies flows from earlier eruptions of Montaña Rajada to the east. The eruption was typical of many of its type in that it began and ended with eruptions of degassed lava, with an explosive phase in the middle. The path crosses ground covered by a veneer of reddish pumiceous blocks (produced by explosions that disrupted lava domes emplaced during the late middle phase of the eruption), overlying the main unit of white phonolitic pumice lapilli ejected during the main explosive phase.

Stop 6.12 Valley between Montaña Blanca and Montaña Rajada   About ten minutes' walk from the road and just past a group of large obsidian blocks, which have rolled down from the flanks of El Culatón, the path passes through a small cut, in the walls of which are exposed a bed of partly welded obsidian blocks underlying a reddish-black glassy deposit of banded lava-like rock. These rocks are analogous to the welded spatter commonly seen in basaltic spatter cones, but the much lower emplacement temperature of phonolite magmas means that such welding can occur only at very high eruption rates. North of this cut, the track passes up the west side of the valley between Montaña Blanca and Montaña

Figure 6.8 Couleé of phonolitic lava on the flank of Montaña Rajada, with characteristic arcuate ridges on the flow top.

Rajada. The top and flanks of Montaña Rajada are covered by phonolite coulées, with good examples of the surface structures characteristic of these viscous lava flows (Fig. 6.8). At the northern end of this valley, 40 minutes' walk from the road, older phonolitic lavas are exposed locally beneath a thick cover of phonolitic pumice. This pumice belongs to the explosive phase of the Montaña Blanca eruption. The thickest deposits of pumice from the Montaña Blanca eruption are in a narrow strip to the northeast, carried there by strong southwesterly winds at high altitude.

Stop 6.13 Lava balls from the most recent summit eruption   Continuing north, a marked bend to the west occurs at a crag topped by a large fragment of a black glassy accretionary lava ball. Similar lava balls occur farther down the slope to the northeast, lying on top of Montaña Blanca pumice. These lava balls were erupted during the most recent eruption at the summit of Teide; some may be fragments that rolled downhill from the front of the lava flows (Fig. 6.9) rather than being derived directly from the summit vent.

To the northeast of this viewpoint is a panorama including the top of

157

Figure 6.9 Phonolite lava flows of the most recent summit eruption, where they have flowed down the eastern flank of Pico del Teide, on top of pumice of the Montaña Blanca eruption. Note the lava balls and fragments littering slope below lava flows.

the La Orotava Valley and the Dorsal de La Esperanza to the east. The eastern wall of the Caldera is also visible, formed by the well bedded sequence of the Diego Hernández formation. To the north is the Fortaleza escarpment, at the north end of the Tigaiga Massif. The upper part of the escarpment is formed by welded phonolitic pyroclastic rocks. These are possibly equivalent to the Diego Hernández formation and they overlie an older sequence of lavas that may be part of the Lower Group of Las Cañadas rocks. Between the Fortaleza escarpment and the base of Montaña Blanca are several basaltic scoria cones. A similar chain of scoria cones can also be seen to the east, notably the sharp peak of Montaña Mostaza.

Stop 6.14 North side of Montaña Blanca The route continues up the north side of Montaña Blanca, following a zig-zag path. At about 3 km from the road, a sharp hairpin bend in the track passes through a cut that exposes the deposits of the explosive phase of the Montaña Blanca eruption. The deposits contain a mixture of well expanded highly vesiculated pumice (70% of the deposit) and low-vesicularity phonolitic blocks (30%; these possibly represent fragments of an early phonolite dome blown to pieces during the explosions), with very rare small lithic fragments. Notable features are the different sizes of the clast types of different density (size inversely correlated with density) and the lack of fine-grain material. These are characteristic features of airfall deposits from convecting Plinian eruption columns. The lack of bedding indicates that, at the time of eruption of this deposit, the eruption column had reached a steady-state condition, with uniform high eruption rate. A little farther on, the track passes through a field of black lava balls erupted in the most recent summit eruption of the Teide volcano: these feature in the tourist guides as Los Huevos del Teide ("the eggs of Teide"). Close inspection reveals that the accretionary lava balls contain feldspar and pyroxene phenocrysts; they are of phonolitic composition and the dark colour is that of the fresh glassy groundmass.

After a 4 km walk (typically taking about 2¼ hours because of the climb and high altitude), the foot of the steep path up to the top of Teide is reached, at the toe of one of the black phonolitic lava flows. The path up Teide is a very serious ascent and will take all day, or can be spread over an evening and a morning, staying overnight at a mountain refugio part way up (see Rochford). The flows of the most recent eruption of Teide are very dark obsidianic lavas (Las Lavas Negras) that stand out against the pale older rocks. The path to the top of Montaña Blanca turns left along the side of one of these flows.

Stop 6.15 Summit of Montaña Blanca   The summit of Montaña Blanca is mantled in two distinct explosive deposits of the 70 BC eruption sequence. The earlier one is a welded obsidian block and spatter deposit, with spectacular flow banding, whereas the later deposit is a very weakly welded white pumice block deposit; the lesser degree of welding probably reflects declining eruption rate at the end of the explosive phase of the eruption. No vents are visible on top of Montaña Blanca, which is probably at least in part a dome or cryptodome, although some vents may be buried under the pumice. The vents for the lava flows formed in the later stages of the eruption form an east–west-trending line or fissure along the south slope of the dome. Red altered blocks ejected during this last stage of the eruption litter the ground.

The summit of Montaña Blanca is a superb viewpoint for the Las Cañadas edifice sequences in the walls of the Caldera (Figs 6.6, 6.10). The unconformities between and within the different formations are particularly notable features. Several of these show deep palaeo-valleys; it seems that the flanks of the Las Cañadas edifice were deeply eroded between the major eruptions. Barranco-filling eruptive units are common, especially in the Guajara formation west of Montaña Guajara.

The return route to the road is identical to the outward route but, being down hill, takes much less time.

The road route continues westwards towards Roques de García. If time is available, a short stop may be made at El Tabonal, 4 km west of the foot of the Montaña Blanca path, to view phonolitic lavas on the caldera floor and a particularly spectacular palaeo-barranco exposed in the south wall, within the Guajara formation (Fig. 6.10).

The road passes the lower station of the cable car, which provides the easiest access to the summit region of Pico del Teide. The cable car opens around 09.00 h on most days (it is often closed because of high winds) and unless you are there early you can expect a queue. As discussed in Chapter 2, you are also most likely to enjoy good views early in the morning, so if you do want to go up Teide it is best to head directly for the cable car station in the morning. The cable car ascends above a field of brown weathered phonolitic lava flows erupted from the second of three craters that have successively formed the summit of Teide, but the best aspect of the journey up is the ever-expanding view of the Caldera. The summit itself is normally closed, although a path around it is sometimes open. The highest point, a small crumbling cone commonly thought to have formed in 1492 but most probably prehistoric, is somewhat disappointing. However, in clear weather the views of the Caldera, the slopes of the island beyond (this is really the place to view the components of Tenerife as shown in Fig. 6.1) and of the other islands can be well worth the journey.

Roques de García
The peculiarly shaped Roques de García are one of the most famous features of the Caldera. They are a favourite stop for coach tours, with a large but often crowded car-park and a group of paths; stick to these paths,

Figure 6.10 South wall of the Caldera de las Cañadas from El Tabonal, showing phonolite lavas and ignimbrites of the Guajara formation. Note irregular contacts between flow units, indicating emplacement in palaeo-valleys eroded on the slopes of the Las Cañadas volcano.

because much of the area around the Roques is covered by protected wild plants.

Stop 6.16 Llano de Ucanca overlook   The best views of the Roques and of the surrounding area are to be obtained by following the well defined path north for about five minutes, to a gap in the crags, through which the western part of the caldera floor and the Llano de Ucanca can be seen. The latter is the largest of the Cañadas, a flat plain covered by pumice and other debris eroded from the cliffs to the south and trapped by thick brown flows of phonolite and basalt that originated from Teide and Pico Viejo. At various times, ephemeral lakes have formed on its surface, and sediments deposited in these form brilliant white patches in the distance.

The Roques de García are formed by a spur of Lower Group rocks that extend out from the wall of the Caldera, dividing the floor into higher eastern and lower western halves. They are made up of an eastward-dipping sequence of lavas (perhaps derived from a volcanic centre to the west), breccias and epiclastic sediments, cut by many inclined phonolite sheet intrusions and showing moderate hydrothermal alteration. They are the oldest rocks exposed anywhere in this central part of the Las Cañadas edifice; they may represent a spur of rock between separate lateral-collapse scars, a remnant of rock left by erosional enlargement of collapse scars, or a ridge of rock left between ancient collapse calderas to east and west.

To the east of Roques de García, the steep western face of Montaña Guajara rises above the floor of the Caldera (Fig. 6.11). Its cliffs are formed by thick phonolitic lavas and welded pyroclastic flows of the Guajara formation. These can be traced to the west for a short distance (Fig. 6.6), but thin rapidly until they form only a cap on the thinner flows of the Ucanca formation; the unconformity between the two is clearly visible at the southwestern corner of Montaña Guajara. It may also be a palaeo-valley, but is not as well developed as the one enclosing the Diego Hernández formation.

Los Azulejos and the western part of the Caldera de las Cañadas
Stop 6.17 Los Azulejos   The less energetic should continue south from the Roques de García car-park for about 1 km, to a cutting through brilliant green rocks: Los Azulejos. These are a group of crags between the Roques de García and the main Las Cañadas wall, composed of breccias and sediments, similar to those at Roques de García but affected by intense hydrothermal alteration. A group of hot springs must have at one time been located above this site, fed by the hot water that circulated through these rocks. The green colour is attributable to the presence of chlorite and iron-rich clay minerals.

Figure 6.11 Montaña Guajara viewed from Roques de García, showing thin-bedded ignimbrites passing up into 80 m thick rheomorphic ignimbrites near top of cliff.

Stop 6.18 Llano de Ucanca: view to Roques de García   West from Los Azulejos, the road descends onto the Llano de Ucanca and continues west along the foot of the cliffs. The rim of the Caldera is formed here by Ucanca formation rocks over thick and varied sequences of Lower Group lavas, pyroclastic rocks and sediments, all cut by many northward-dipping inclined sheet intrusions and dykes. These are mainly phonolitic and often thick, forming prominent walls and slabs in the cliff. It is often worth stopping at a small parking place about 2 km from Los Azulejos, just beyond a small roadcut, for a fine view back to the Roques de García and of Teide.

The road junction about 5 km to the west offers two alternatives: either continue south through the Boca de Tauce on the C-821 or turn north on the C-823. The former is the route to the south coast, via Vilaflor; the latter is the longer route back to the north coast, via Santiago del Teide and Icod. Unless you want a first look at the western side of the island, it is better to return to La Orotava by retracing your route on the C-821.

If time permits, it is worth following the C-823 for a few kilometres, to where it crosses the fresh fields of basalt lava flows erupted in the recent past from Pico Viejo, on the western side of Teide. The youngest of these were erupted in 1798 and are still black and almost completely barren.

Stop 6.19 View of Pico Viejo   In the middle of these flows, a parking area on the right of the road offers a view of Pico Viejo to the northeast; from here, the line of vents from which these flows were erupted can be traced almost up to the summit of Pico Viejo. The flow tops exhibit many fine examples of the rough, clinkery **aa** flow morphology and a few areas of the smoother **pahoehoe** flow type (be very careful if you venture on the former in particular, as they make for very difficult walking).

The southern exit from the Caldera is a winding mountain road. A few parking places in the first 3 km offer opportunities to examine the steep mountain and valley walls formed by the south- and west-dipping sequences of Lower Group and Ucanca formation rocks (lavas, sediments and ignimbrites) where they were deposited on the outer slopes of the Las Cañadas volcano; and, in clear weather, views west and southwest to La Gomera and El Hierro. Lower down, the road descends through the pine forest, and exposures become more limited. Furthermore, there are very few places where it is safe to stop, and it is therefore better to examine the rocks deposited farther down slope from the volcano in the Bandas del Sur, closer to the coast.

The Dorsal de La Esperanza

This route follows the C-824 from La Laguna to the junction with the C-821 at the eastern end of the Caldera (El Portillo): it is best to follow the route in this direction (especially in the morning) as it then affords many views of Pico del Teide.

As discussed in the introduction, the Dorsal de La Esperanza is presently occupied by the active northeastern rift zone of Tenerife, and several recent fissure eruptions have taken place along it. The ends of the ridge, near La Laguna and on the eastern rim of the Caldera respectively, are mostly covered by recent basaltic lava flows and scoria cones. In the middle are the eroded remains of two large volcanoes and the headwalls of the collapse scars left by lateral collapses of these volcanoes, which today form the La Orotava and Güímar valleys (Fig. 6.2; see also Fig. 1.4).

From the junction with the TF-5 (Autopista del Norte) on the outskirts of La Laguna, the C-824 first passes through agricultural land and the small town of La Esperanza.[*] The road then climbs through the laurel and pine forests onto a group of scoria cones that form the eastern end of the Dorsal. A viewpoint at Pico de Las Flores offers a view back over the remarkably flat Vega de La Laguna to the rugged crest of Anaga. The contrast between the deeply eroded rocks of Anaga (the oldest part of Tenerife) and the much younger and little-eroded rocks of the Dorsal is clear at this point.

Stop 6.20 Risco Negro   At 18 km from La Laguna, a viewpoint at Risco Negro offers the first clear view of Teide, rising beyond the northern flank of the Dorsal. Views along much of the rest of this section of the Dorsal are obscured by pine forests. As the route climbs to 1700 m altitude, however, the pine forest begins to thin and the road continues along a narrow ridge, with especially steep slopes to the south. These form the eastern end of the Güímar collapse structure. This structure (perhaps 1.2 million years old) has been partly filled by younger lavas and then re-excavated and enlarged by erosion. A narrow road (TF-4132), descends into it at around the 25 km mark, but the best views of Güímar are to be found farther west along the ridge.

Stop 6.21 Risco Yesa   About 1 km farther on, the broad La Orotava Valley comes into view. In good weather, a viewpoint at Risco Yesa (27.5 km from La Laguna) offers an excellent view of the eastern part of this collapse structure. In contrast to the floor of the valley to the west, which has been

---

[*] It is worth keeping note of the distance travelled from the start of the TF-824 in La Laguna, because some of the stopping points are less than obvious.

partly filled by younger lavas, this eastern part has been enlarged by the growth of a major barranco system, centred on the Barranco de Pedro Gil. The sequence exposed in the cliffs of this barranco consists primarily of lavas and breccias of the Güímar volcano, dipping north and east, and cut by mainly northeast-trending dyke intrusions that can be traced across the slopes. Both the dip of the volcanic rocks and the trend of the dykes are radial to the centre of the Güímar volcano, lying 2 km to the southwest.

Stop 6.22 Montaña Ayosa   Less than 1 km farther on (28.2 km), a second and larger parking bay at the head of Barranco de la Madre is located at the centre of a roadcut excavated into Montaña Ayosa. This peak is formed by thick trachybasaltic lava flows, quite distinct from the basaltic rocks of the Güímar volcano, filling a palaeo-valley that extended down hill from the northwest and whose sloping sides can be seen at either end of the roadcut. The flows filling the palaeo-valley must have therefore originated from a volcano that occupied the present position of the La Orotava collapse structure and of course postdated the Güímar volcano.

Stop 6.23 Caldera de Pedro Gil   The road continues down hill for about 0.5 km, then re-enters the pine forest and crosses the crest of the narrow ridge separating the La Orotava Valley from a steep-walled, almost completely enclosed pit about 2 km across and several hundred metres deep. The road continues around this pit to La Crucita (30 km), where it is possible to park. A gravel track descends to the south, and a cobble path descends to the north from a point about 50 m to the west.

The Caldera de Pedro Gil is best seen from a few hundred metres south of the parking place; follow the gravel track to the first hairpin bend. The walls are formed by lavas and breccias that everywhere dip away from the pit and are cut by many radial dykes. This is the centre of the Güímar volcano. However, it is not a simple crater, as reconstruction of the volcano from the dips of the lavas on its outer flanks indicates that its summit was originally as much as 1 km higher. Rather, the collapse of the south flank of the volcano and subsequent erosion have excavated its hydrothermally altered core to produce the pit, whereas the surrounding rocks have remained intact, buttressed by the dykes.

Subsequently, eruption of lavas and scoria along the Dorsal de La Esperanza rift zone has partly filled the Caldera de Pedro Gil. The feeder vent for one of these eruptions can be seen at the hairpin, in the form of a feeder dyke that expands upwards into a spatter-filled pit, now exposed by erosion of the lapilli beds that formed its walls. The vent of the most recent eruption can be seen at the mouth of the Caldera de Pedro Gil: a fresh black scoria cone that formed in 1705, blanketed the surrounding area (Las

Arenas Negras: "black sands") in lapilli and fed a lava flow that descended into the Güímar Valley.

Stop 6.24 Curvas del Pastel   West of La Crucita, the road climbs out of the pine forest and into an area mainly covered by relatively recent (although prehistoric) beds of lapilli and scoria erupted from vents on the northeast rift zone. At around 32 km, the road climbs steeply through a series of hairpin bends cut into a striking alternation of black lapilli and pale phonolitic pumice and ash beds (Fig. 6.12): the Curvas del Pastel. Parking bays are available (and often occupied by tourist buses). The lapilli beds were erupted from nearby vents, but the pumice beds represent distal fallout from sub-Plinian eruptions, most probably of the early stages of growth of the Teide–Pico Viejo complex within the Caldera. The well sorted nature of the pumice, nearly all of which is around 1–2 cm across, is characteristic of airfall deposits, as is the lack of bedding in most of the deposits. The development of bedding in places is most probably the result of reworking of pumice from farther up hill at the end of, or after, the eruptions.

Besides examining the main pumice beds, it is worth walking back down hill to the first sharp bend. Here, the roadcut on the east side of the road is mainly in oxidized red lapilli and scoria, but steeply inclined fissures within it are filled with altered yellowish pumice. This indicates that the fissures opened up during the eruption of the pumice, as they would not have remained open for long; they were probably produced by earthquakes associated with the explosive eruption that deposited the pumice. This same bend also offers a good view down into the La Orotava Valley, especially towards the west, where the very straight western sidewall is clearly visible (Fig. 6.13). These very linear sidewalls are characteristic of valleys produced by lateral collapses; although the eastern wall of the La Orotava collapse structure has been greatly modified by erosion, much of the overall shape of this western wall is still intact.

Almost directly below this viewpoint, erosion of the thick sequence of lapilli and ash that has accumulated in this part of the valley has produced an area of gullies in which a similar alternation of black lapilli and white phonolitic pumice is exposed, but on a larger scale. At least three pumice beds are present, the largest of which is several metres thick; these indicate that in the relatively recent past Teide has produced explosive eruptions significantly larger than the Montaña Blanca eruption. This deposited only several centimetres of pumice at comparable distances from the vents.

Stop 6.25 Güímar Valley overlook   At about 34 km from La Laguna, the road crosses over onto the south side of the Dorsal de La Esperanza again; there is another convenient parking bay from which to view the Güímar

Figure 6.12   Interbedded white phonolitic pumice and black basaltic scoria at Curvas del Pastel.

Figure 6.13 Western side of the La Orotava Valley from Curvas del Pastel (stop 6.24). The steep, notably straight western wall of the valley runs down slope from the Fortaleza escarpment (see stop 6.13) just visible to the north of Teide on the north side of the Caldera de las Cañadas.

Valley over pine-covered slopes. As noted previously, the Güímar Valley has been partly filled and then re-excavated since its original formation. Thus, part of the well exposed lava sequence that forms its western wall, rising to the ridge of Izaña (at some 2400 m in elevation), actually postdates the Güímar collapse and is part of the La Orotava volcano.

Stop 6.26 North side of Izaña   These younger rocks can be seen in the stretch of road from km-35 to km-37.5, which runs below the steep northern face of the Izaña Ridge. The buildings on the ridge are telescopes of the Canarian Astrophysical Observatory, which is mainly devoted to observations of the Sun. The thick, trachybasaltic to phonolitic lavas that form the ridge are an excellent stable foundation for these telescopes, although in places they are interbedded with pumice beds (best seen around the km-37 mark). A notable feature of this sequence of lavas is that it dips markedly to the south-southeast; this and the presence of palaeo-channels, filled with reworked pumice and lapilli (Fig. 6.14), running in the same direction, indicate that the summit of the volcano concerned lay to the north, in what is now the La Orotava Valley. Reconstruction of this volcano, based upon the point of intersection of the dykes that cut this sequence and also occur at other places of the La Orotava Valley, suggests that its summit lay 3–5 km north of Izaña and was over 3000 m high.

The western end of the Izaña ridge, at about the 37.5 km mark, is also very steep. A junction at this point marks the beginning of the road to the observatory. This is closed to the public, but it is worth going up this road for about 1 km to gain a view out to the west, including the rim of the Caldera and Teide itself.

Figure 6.14 Well bedded channel-filling pumice and lapilli beds exposed in roadcut on the north side of the Izaña Ridge (stop 6.26). These are not primary deposits, but were reworked down slope on the south side of a volcano that occupied the area La Orotava Valley before undergoing lateral collapse to the north.

To the southwest of this viewpoint, almost on the crest of the slope that descends to the south coast of the island, is a group of very fresh scoria cones and small basaltic lava flows, startlingly black against the weathered pumice plain. These were erupted in late 1704, in the first of the three eruptions that occurred over little more than 18 months and which culminated in the destruction of Garachico (see stop 6.8, p. 150). They were fed by a single dyke emplaced along the south side of the northeast rift zone – hence the cone alignment along this trend – which may have subsequently propagated eastwards, deep underground, to feed the single vent in the mouth of the Caldera de Pedro Gil (stop 6.23, p. 165). These cones can be reached by returning to the main road and following the dirt track that starts about 50 m east of the junction and winds its way south over the pumice plain for about 3 km before passing close to these cones. Much of this area is restricted, however, so be careful not to lose your way.

The main road continues west through the field of scoria cones and lava flows for another 5 km, before reaching the road junction at El Portillo. It is then possible either to descend the C-821 through the La Orotava Valley to Puerto de la Cruz, or to continue south through the Caldera.

The Bandas del Sur

The Bandas del Sur is the dry southern part of Tenerife, between the Güímar Valley and the west coast. It is a gently inclined slope, extending from sea level up to around the 500 m contour, cut by many steep-sided barrancos that greatly hindered access to the area. Until 30 years ago it was the poorest and most remote part of the island, but since then it has become a centre of mass tourism. In the process, the old fields have been abandoned and much of the coastal area concreted over. Nevertheless, it offers opportunities to examine the distal deposits of the Las Cañadas volcano, consisting of a variety of phonolitic pyroclastic rocks overlying a rather monotonous sequence of basalts. The latter mostly correspond to the Lower Group sequences exposed in the Caldera, whereas the pyroclastic rocks correspond to the Upper Group. Several correlations have been proposed between individual eruptive units and the formations of the Upper Group (see Table 6.1), but proving these is problematic because the forested zone at 1000–1700 m elevation to the north is both poorly exposed and deeply eroded, preventing the direct tracing of particular beds down slope from the rim of the Caldera, visible on the horizon to the north in clear weather, to the Bandas del Sur.

Many of the best exposures in the Bandas del Sur are in roadcuts. Best of all are those along the Autopista del Sur, but do not be tempted to stop, as this is against the law. However, it is convenient to stop along some of the minor roads off the motorway, but beware of local traffic while examining the following (described in an east-to-west sequence).

Stop 6.27 Fasnia   The road to Fasnia (TF-6133) passes beneath the autopista at the junction and climbs up and around an old inter-ignimbrite scoria cone. Pumice beds drape this cone and are well exposed on its southern face 3 km from the junction in an abandoned roadcut. They consist of well sorted airfall pumices in beds with cross bedding and some sorting, suggesting that the original deposit was reworked down the slope of the cone. Farther on, in a roadcut and abandoned quarry on the edge of Fasnia (5 km from the junction), there are extensive exposures of indurated but originally unwelded ignimbrite. Most of the pumices in this thick deposit have been zeolitized and intensely altered, but some large pumices are still green (indicative of their alkali- and iron-rich composition) and relatively fresh.

Stop 6.28 Tajao   On turning off from the autopista, the road to Tajao (which lies on the coast to the south) is to the left at the T-junction at the end of the slip road. This road descends to the village through a roadcut; after parking on the right at the edge of the village, walk back up through the roadcut. **Beware of traffic**. The roadcut exposes a sequence of pumice-rich unwelded ignimbrites with well developed pumices and some lithic rich concentrations. Intercalated Plinian airfall pumice beds with excellent sorting of clasts are also present. The contacts between the units are erosive in places. Did the ignimbrites fill channels eroded before or during eruptions, or are they reworked secondary ignimbrites such as those produced by rapid erosion of still-hot primary ignimbrites after the 1990 eruption of Pinatubo in the Philippines? These deposits raise the interesting question of the extent to which these deposits have been reworked onto the coastal plain by rapid erosion and re-deposition after the primary eruptions, as occurred on a very large scale in the case of the 1990 Pinatubo deposits.

Stop 6.29 El Médano   El Médano is a tourist resort of little architectural merit, but after parking in the centre of the town and walking down to the beach, continue west around the bay towards the prominent, steep-sided eroded volcanic vent of Punta Roja. From where the buildings run out, the low bluffs at the back of the beach consist of initially unwelded but strongly indurated pumice-rich ignimbrites, interbedded with cross-bedded beach sands containing the remnants of shell fragments. The beach sands contain vertical bands of hard cemented sandstone, often with eroded-out cores, and indurated masses of sandstone that have resisted erosion, whereas the sands around them have been washed away, to produce strange mushroom-like shapes. These features seem to have been produced when the hot ignimbrites flowed over wet beaches, boiling the water in the beach sands. The hot water and steam dissolved many shell fragments and was then ejected upwards through fractures. The cooling water reacted with the sand in the walls of the fractures to produce the vertical bands, or soaked into the sands to produce the mushroom-shape masses. Besides providing an unusual example of ignimbrite–water interaction (which is usually much more explosive), these rocks provide a very useful sea-level indicator. Since they were deposited in the middle Pleistocene, perhaps 800 000 years ago if the correlation of the sequence of ignimbrites in this area with the Ucanca formation is correct, their occurrence at sea level today indicates that the island of Tenerife is now remarkably stable and is not subsiding, unlike, for example, the Hawaiian Islands, as discussed in Chapter 1.

Chapter 7

La Gomera

Introduction

La Gomera, the smallest of the Canaries after El Hierro, is a shield volcano, $380\,km^2$ in area and 25 km in diameter, with a maximum height of 1487 m in the central sector at Garajonay. Although it developed very close to Tenerife (22 km), they are completely independent island-volcanoes, as is shown by their different ages and the 1500 m depth of ocean between the islands. La Gomera developed predominantly in the Miocene, with lesser post-erosive volcanism in the lower Pliocene. Unlike in the other Canaries, volcanism ceased completely here nearly 4 million years ago.

Whereas juvenile oceanic islands such as La Palma and El Hierro show excellent examples of constructive volcanic features, La Gomera shows the best examples in the Canaries of erosive features and landscapes, similar to those present in Kauai, the oldest of the Hawaiian Islands. The topography of La Gomera is characterized by an eroded central plateau of horizontal lavas (similar to those described in La Palma and El Hierro) from which a network of deep, frequently amphitheatre-headed, barrancos radiates (Fig. 7.1), very similar to those of the Hawaiian island of Oahu. Many lava plugs and domes (roques or fortalezas) outcrop in the island. The characteristic cinder cones are absent in La Gomera (the only well preserved volcanic cone is La Caldera, at the southern coast, dated at 4.2 million years old). Precipitous vertical cliffs, up to 800–900 m high, generally form the coast. Coastal rockfalls are frequent and scarps common, the last rockfall occurring in 1949 at the mouth of the Barranco de La Majona on the northeast coast (Fig. 7.1). The few predominantly basaltic pebble beaches are located at the mouth of the barrancos. These features correspond to an old and deeply eroded island without Quaternary volcanism.

Blumenthal (1961), Bravo (1964), Hausen (1971), Cendrero (1971) and Cubas (1978) have studied the geology of La Gomera. Radiometric dating has been carried out by Abdel-Monem et al. (1971), Féraud et al. (1985) and Cantagrel et al. (1984). A 1:25 000 geological map of the island has been

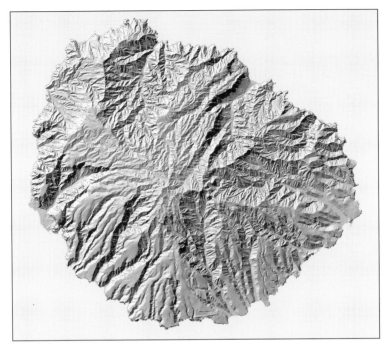

Figure 7.1 Shaded relief image of La Gomera (image GRAFCAN).

published by the Instituto Tecnológico y GeoMinero de España (available from igme@igme.es).

The submarine volcano (the pre-shield stage)
The oldest geological formation, comprising the uplifted seamount and associated plutonic rocks, outcrops in the north and northwest of the island (Fig. 7.2). This unit, very similar to those described in La Palma and Fuerteventura (also known as the Basal Complex), represents the pre-shield stage of growth of the island, mainly the seamount stage. As mentioned in the chapters on Fuerteventura and La Palma, it is formed of submarine (basaltic and trachytic) pillow lavas and tuffs and plutonic rocks.

In La Gomera the main part of the outcrops is gabbros intruded by an extremely dense dyke swarm, which accounts for up to 90 per cent of the unit. These plutonics intrude into the older submarine rocks. The radiometric ages published (12–20 million years ago) are of very low reliability.

The shield-stage volcanism: the Miocene basalts and trachy-phonolitic complex
The initial subaerial volcanics, overlying the seamount on a general discordance, mark the onset of the island shield stage. Basaltic lavas (mainly

173

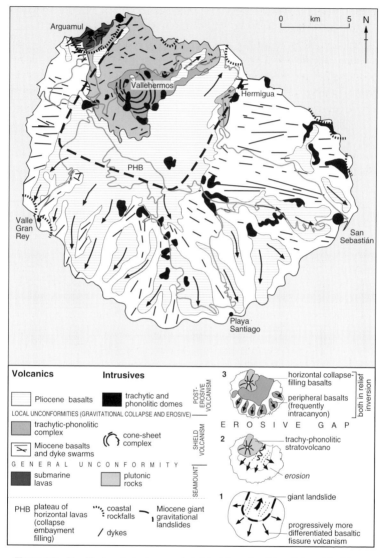

Figure 7.2 Simplified geological map and sketch of the subaerial evolution of the island.

thin pahoehoe flows) and pyroclastics cut by a dense dyke swarm form this 250 m-thick lower subaerial unit, which is best exposed in the area of Tazo and Alojera to the northwest and in the valley of Hermigua (see Figs 7.1, 7.2). These outcrops represent the remains of a large volcano, the north flank of which apparently collapsed.

Volcanic activity continued uninterrupted after the collapse, as occurred in the Garafía and Taburiente volcanoes in La Palma and the El Golfo volcano in El Hierro developing a central volcano with terminal differentiated rocks. This volcano subsequently collapsed towards the north, its remains cropping out in the caldera-type depression north of Vallehermoso. The remains comprise lavas (trachytic and phonolitic) and intrusives, the latter forming radial swarms and cone sheets. The analysis of their dip angles and distribution suggests that the cone sheet originated from an ascending magma body located some 1350 m below sea level.

The post-erosional Pliocene basalts
The most recent basaltic unit is separated from the old shield volcanoes by a marked unconformity in the western part of the island, whereas in the remaining parts the young basalts rest in apparent conformity. The age of the main volume of these post-erosional basalts is lower Pliocene.

Between the shield and the post-erosional stages, both a northward-trending giant landslide (Figs 7.1, 7.2) and an important interruption of volcanic activity (up to 4 million years) may have occurred. The filling of the collapse embayment gave place to a plateau of horizontal lavas, in clear erosive and angular discordance with the underlying Miocene basalts. This process is very similar to the formation of the central plateau of horizontal lavas in the northern shield of La Palma and in El Hierro. These geological units form the main aquifers of the three islands. In La Gomera, it is clearly marked by a line of springs at the discordant contact with the Miocene, which are the main water resource of the island.

Many eruptive vents opened during the Pliocene on the flanks of the old shield (peripheral volcanism), emitting a thick (500–1000 m) section of basaltic lava flows and pyroclastics. The Pliocene lava plateau is encircled and often intruded by trachytic and phonolitic domes, among the most conspicuous and spectacular volcanic features of the island. The Pliocene volcanic activity ended at about 4.2 million years ago, the age of the only recognizable cinder cone (La Caldera) located in the south of the island (see Figs 7.1, 7.2).

Logistics on La Gomera

La Gomera is well connected by fast ferries (40 minutes) with Tenerife (between Los Cristianos and San Sebastián). There are two main tourist resorts: Playa Santiago on the south coast and Valle Gran Rey to the southwest. Hotels and pensiones are also available in San Sebastián, Hermigua and Vallehermoso. The rural tourism houses are an interesting alternative (www.aecan.com).

The weather is similar to La Palma and El Hierro, frequently with a sea of clouds on the northern flanks, while the southern coasts are almost constantly clear and sunny. Keep away from the unstable coastal cliff edges and do not enter the ravines on rainy days.

There are about 16 natural parks and reserves; the most stringent restrictions (www.gobcan.es/medioambiente) are applied in Garajonay National Park, which covers the central plateau of the island (Fig. 7.3).

There are three main roads (generally winding and slow), at the northern and south coasts and over the central highlands. Many tracks can be used by four-wheel-drive vehicles, and most trails and paths are well maintained.

Road and trail maps are abundant. Precise topographic maps (1:50 000 to 1:5000) and aerial photos, showing every road and track, are available from the Instituto Geográfico Nacional (www.cnig.es) and the Canarian government (www.grafcan.es). Geological maps of La Gomera have been published by the Instituto Geológico Nacional (igme@igme.es).

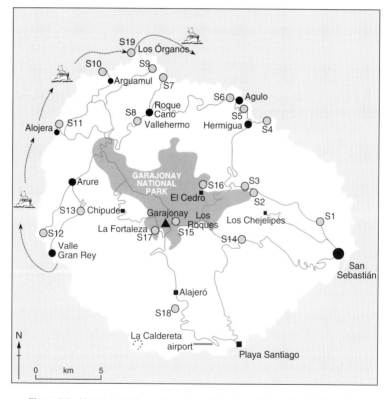

Figure 7.3 Map showing the main roads and the stops of the geological itineraries.

Itineraries

Three itineraries are suggested:

- along the western and northern roads: the Miocene seamount and shield volcanoes
- on the central and southern roads: the Pliocene, post-erosional volcanics
- a boat trip around the island: the coastal features and barrancos.

Itinerary in the north: the seamount and the Miocene basalts
This itinerary will take at least a full day, starting in the capital, San Sebastián, and driving on roads and well kept tracks.

Stop 7.1 Quarry at El Molinito The northern road to Hermigua and Vallehermoso (TF-711) leaves San Sebastián crossing lavas and pyroclasts of the Miocene shield and scree deposits. On arriving at a sharp bend (after 3 km) with a bus stop to the right, turn left into a track that runs up hill into a quarry located in a fossilized cinder cone. The weathered and cemented red pyroclasts can be cut into blocks and were used to construct the principal buildings in the capital town.

Stop 7.2 Entrance to the main tunnel The road climbs the flank of the Miocene shield, crossing a monotonous section of weathered basaltic lava flows, cinder beds and dykes about 9 million years old. Some of these dykes are several kilometres long and form very conspicuous wall-like features, locally known as taparuchas, the longest of which can be observed and photographed on the road from San Sebastián to Los Chejelipes (Figs 7.3, 7.4).

The road passes through several short tunnels. Stop before entering a long tunnel at the top, which opens onto the northern flank of the shield and into the valley of Hermigua. From this point a general view of the caldera-type (Oahu-type) Barranco de La Villa can be observed. The barranco is carved on Miocene basaltic flows, dipping towards the sea. A relatively dense dyke swarm cuts into the lavas. The homogeneity of this formation and the ages tightly grouped around 9 million years ago are in agreement with the relatively short, continuous and highly productive volcanic activity of the shield stage of oceanic islands.

The southern interfluve is made up of Pliocene lavas in inverted relief topping the Miocene basalts. On clear days, the phonolitic domes of La Zarcita and Ojila can be seen to the southwest at the head of the Barranco de Las Lajas, a tributary of Barranco de La Villa.

Figure 7.4 (a) Thick (4–5 m) basaltic dyke, several kilometres long, intruding the Miocene basalts in Los Chejelipes, in the Barranco de La Villa near San Sebastián. These long wall-like dykes are known locally as taparuchas. **(b, opposite)** Basaltic dyke in one of the oldest outcrops of the Miocene basaltic shield, in the valley of Taguluche.

Stop 7.3 Main tunnel exit Park at the entrance to a short track to the right upon leaving the tunnel. The scenery, and often the weather, change considerably as the 500 m tunnel crosses a sharp 950 m-high ridge blocking the trade winds. The valley of Hermigua is generally more humid and cloudy.

The view of the left (west) wall of the valley shows the Pliocene basalts resting unconformably on top of the Miocene shield. These younger lavas filled one of the main barrancos that opened the Miocene landslide concavity to the sea and now appear in relief inversion. Conversely, the eastern wall is composed entirely of Miocene basalts. Erosion and vegetation have created a landscape resembling the Nuuanu Pali in the Koolau volcano (Oahu) or the Na Pali coast in Kauai.

The winding road descends, offering closer views of the valley walls and the relationships between the Miocene and Pliocene volcanoes. Weathering of the lavas increases, as does the number of dykes, as the road runs deeper into the Miocene shield.

Past km-19 the road enters the plutonic rocks (gabbros, pyroxenites) of the seamount stage (see Fig. 7.2). A road to the right crosses the barranco to Las Cabezadas and follows the eastern footwall of the barranco, where interesting outcrops of the submarine volcanics and plutonic intrusions can be seen.

179

Figure 7.5 **(a)** The mouth of the Barranco de Hermigua from the top of the western wall at Los Risquetes. The cliffs behind the alluvial flat at the bottom of the barranco are made of the oldest lavas of the Miocene basaltic shield. **(b, opposite)** View from the sea of the south coast showing the flat-bedded valley of the Barranco de Chinguarime, east of Playa de Santiago, with the walls made of Pliocene (peripheral) basalts.

Stop 7.4 Playa de Hermigua   Follow the road to Playa de Hermigua, at the mouth of the valley. The last stretch of the Barranco de Hermigua shows the typical features of the mature ravines of La Gomera (Fig. 7.5a), with smooth valley-side slopes formed by scree deposits (usually terraced for farming) and wide flat valley bottom, filled with alluvial deposits (now occupied by banana plantations). Follow the coast road to the east and park where the road has been cut by marine erosion (the road served the dock seen farther east, built in 1923 to ship bananas for export).

On clear days Tenerife can be seen to the north (the silhouette of the Teide volcano will be see at each bend of the north road to Vallehermoso). The wide plateau at the top of the western wall of the Barranco de Hermigua, made up of Pliocene basalts, is visible in the foreground. The cliff to the east is formed by deeply weathered and densely injected basaltic flows, corresponding to the earlier phases of activity of the Miocene shield.

From Llano Campos, 1.5 km to the south, a very steep track requiring a four-wheel-drive vehicle climbs the entire eastern wall of the barranco to the Riscos de Juel (700 m above sea level). A forest track at the top contin-

(a)

ues over Miocene basalts to Enchereda and Las Casetas, at km-7.2 of the TF-711 road, near stop 7.1. This track shows a complete section of the Miocene sequence.

Stop 7.5 Agulo Return to the main road (TF-711) to Agulo and follow the Miocene/Pliocene discordant contact. Agulo lies on the 2 km-wide alcove of a coastal rockfall. Similar rockfalls are frequent at the coasts of La Gomera, La Palma and El Hierro, greatly contributing to the erosive retrogression of the coast. The collapse cuts the final stretch of Barranco de La Palmita above Agulo, leaving a hanging valley that used to include a spectacular waterfall, which has ceased to flow since the construction of the Embalse de Agulo irrigation dam.

Stop 7.6 Exit from the Agulo tunnel The short tunnel just after Agulo crosses a spur of columnar-jointed Pliocene basalts. On the western side of the tunnel the road cuts the first easily recognizable outcrops of the pre-shield plutonic rocks, dark gabbros intruded by a northeast–southwest-trending dyke swarm so extremely dense that the host rock is frequently reduced to small screens between the dykes. On the west wall of the barranco, the Pliocene basalts can be seen lying in clear erosive and angular discordance on top of the pre-shield plutonics. These plutonic rocks form most of the northern part of the island to Vallehermoso.

Stop 7.7 Playa de Vallehermoso Park at Vallehermoso (km-40) to observe and photograph the 200 m-high phonolitic plug of Roque Cano (see Fig. 7.9d). Turn north in the centre of the town to Playa de Vallehermoso (a road signposted Parque Maritimo, TF-712). Walk along the coast to the north to observe interesting outcrops of the pre-shield plutonic rocks and associated dyke swarm.

Stop 7.8 Vallehermoso–El Bailadero: the trachy-phonolitic complex From Vallehermoso, the road climbs to the crossroads at El Bailadero (840 m above sea level), cutting through the trachy-phonolitic complex, which appears to be the deep structure of a large stratovolcano of differentiated (trachytes, phonolites) rocks. This formation is similar to, although smaller than, the Miocene caldera in central Gran Canaria (see Ch. 5).

Deeply weathered trachytic tuffs and breccias, thick trachytic and phonolitic dykes, forming a radial and cone sheet complex, and syenitic intrusions can be observed in the roadcuts.

Stop 7.9 Cumbre de Chigueré At the crossroads at El Bailadero, turn right to the road signposted Alojera and farther down follow a dirt track to the right to Tazo and Arguamul. After a few kilometres there is a track to the right, climbing to the Altos de Chigueré.

This track follows the foot of an elongated spur of sub-horizontal Pliocene basaltic flows filling a barranco carved in the seamount stage plutonics, in a clear example of inversion of relief. The track, routing along the crest of the steep-walled valley of Arguamul, enters the Altos de Chigueré, crossing thick northwest–southeast-trending dykes, some of them trachytic, and ends at the edge of a 560 m-high coastal cliff. From this spectacular viewpoint, impressive views of the northern coast, with precipitous cliffs carved in the pre-shield plutonics, can be seen (best in the afternoon for photography; Fig. 7.6a). To the south, a complete view opens over the valley of Vallehermoso, with the trachy-phonolitic complex on the valley floor. The island of Tenerife and the Teide volcano can be seen in the distance on clear days.

Stop 7.10 Arguamul   Return to the track to Tazo and turn left (north) to Arguamul. The track enters the valley of Arguamul, capped on the western side by another sequence of Pliocene flows filling a former barranco in the pre-shield rocks. The valley is probably a relatively recent (post-Pliocene) collapse feature. The rocks are intensely brecciated and, in some cases, show plastic deformation associated with mass sliding or creeping. From the mirador above the village, screens of dykes folded by the rock drift can be observed (Fig. 7.6b).

From the small village of Arguamul, a path descends to the beach (Playa de Arguamul), where the features described can be observed and photographed.

Stop 7.11 Alojera   Return to the track (only for four-wheel-drive vehicles) signposted Tazo–Alojera, which runs parallel to the Barranco de Tazo to the coast. West of the track, the intensely dyke-injected oldest Miocene basaltic flows appear, topped by inversion relief remnants of Pliocene flows filling former barrancos. The track ends at the fishing village and beach of Alojera, a good place to rest while observing the spectacular cliffs carved in the Miocene shield volcano. Thick light-coloured phonolitic dykes crop on the north cliff of Playa de Alojera (see Fig. 7.11), corresponding to late Miocene intrusions.

Stop 7.12 Arure–La Mérica   Return to the main road, signposted Epina and El Bailadero, to join the main road and continue to Arure. The road runs over an eroded tableland of horizontal Pliocene basalts filling the western part of the Miocene giant landslide. At Arure, take the track to the right, signposted Ermita del Santo, to Taguluche–La Mérica.

Figure 7.6 (a) Dense dyke swarm in the submarine formation at Playa de Arguamul. **(b)** Intensely fractured dykes in the path to Playa de Arguamul. In some outcrops of the valley of Arguamul the dykes are folded, probably because of plastic deformation.

This track affords a spectacular view of the headwall of Barranco de Arure, an alcove carved in the horizontal Pliocene basalts, and follows the crest line of Pliocene basalts. There are impressive views of the valley of Taguluche, carved on the initial phases of the Miocene shield.

The track continues to a site where domestic refuse is usually smouldering (in a protected rural park!). From the top of the crest the best close view of the discordant Miocene–Pliocene basalts can be observed (to the southeast, Fig. 7.7a; best in the afternoon for photography). The track changes to a paved path to the top of Los Llanos de La Mérica (an abandoned farm). Good views of the west coast and the valley of Valle Gran Rey can be enjoyed from the top. Far to the east, the trachytic plug of la Fortaleza de Chipude can be seen.

The path ends at the top of the 600 m-high vertical cliff above the town of Valle Gran Rey. The cliff is the scarp of a large rockfall, the debris talus of which can be seen at its foot.

A well kept paved path descends the sheer vertical cliff to the village of Valle Gran Rey over Pliocene and Miocene lava flows.

Stop 7.13 Valle Gran Rey Back in Arure, take the road down to Valle Gran Rey. Park at the César Manrique viewpoint. This mirador offers the best view of the discordant contact of the Pliocene horizontal basalts and the steeply dipping Miocene flows (Fig. 7.7b). The barranco shows an excellent example of relief inversion (Fig. 7.8). The contact is marked by a line of springs discharging from the main aquifer of the island, stored in the Pliocene basalts, which are horizontal because they filled the collapse embayment (see stops 8.6 and 8.8 for further details). The main springs, channelled to irrigate the banana plantations at the coast, surface at the head of the barranco, in La Vizcaína.

The tourist resort of Valle Gran Rey is situated on alluvial and debris-avalanche talus, on vertical coastal cliffs formed by rockfall at both sides of the mouth of the barranco (see Figs 7.1, 7.2).

The centre and south: the Roques (trachytic domes) and Pliocene basalts
This itinerary is focused on to the observation of the many trachytic and phonolitic domes, and the Pliocene rejuvenation volcanism. It involves a day-long drive along the central and southern roads of the island.

Stop 7.14 Degollada de Peraza viewpoint The central road, signposted Playa Santiago, leaves San Sebastián to the west and climbs towards the central heights of Garajonay (1487 m above sea level). For the first few kilometres the route passes over Miocene basalts and later intracanyon Pliocene flows filling former barrancos. Magnificent views of Tenerife and El Teide can be seen as the road climbs.

Figure 7.7 **(a)** Close view of the discordant contact between the Miocene shield basalts (MB) and the post-erosional horizontal Pliocene basalts (PB). In many outcrops the basal part of the Pliocene volcanics are Strombolian deposits (ST), corresponding to the more explosive initial events. **(b)** General view of the Miocene/Pliocene discordant contact in the eastern wall of the Barranco de Valle Gran Rey.

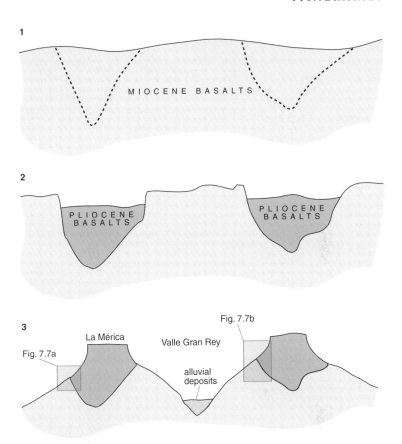

Figure 7.8 The formation by inversion of relief of the barranco of Valle Gran Rey, southwest La Gomera.

Several outcrops of white phonolitic breccias appear at the roadcuts between km-14 and km-15. These ash beds of a nearby explosive eruption form apparent synclinal folds, but they are relief-filling airfall deposits conforming to the previous topography.

Park at the mirador at km-16 for a wide panoramic view to the north of the Barranco de La Laja, excavated on seaward-dipping Miocene basalts, with dykes like walls several kilometres long (the "taparuchas" of stop 7.2). The view to the south shows deep radial barrancos with sharp inter-fluves, flat-bottomed beds and amphitheatre-shape headwalls (Oahu-type barrancos), typical of mature volcanic landscapes.

Stop 7.15 Mirador de Los Roques Follow the road to Arure (TF-713), which cuts thick horizontal flows (collapse embayment filling) of the

Pliocene rejuvenation stage. Past km-18 the road opens onto a view of the Roques, a group of trachytic and phonolitic volcanic domes (Fig. 7.9). The Roques are endogenous domes, grown primarily by expansion from within as concentric flow layers. To the south lies the Roque de Agando (Fig. 7.9a), and to the north the Roques of La Zarcita, Ojila and La Laja (Fig. 7.9b). To the northwest, the thick sequence of horizontal Pliocene basalts can be clearly seen resting discordantly over the Miocene basalts at the head of the Barranco de Benchijigua.

Past the next bend a short paved path leads to the viewpoint affording the best close-ups of the Roques de Ojila and La Zarcita. Farther up hill, there are other miradors with views of the Roque de Agando and the Barranco de Benchijigua.

Stop 7.16 El Cedro The horizontal Pliocene basalts are almost entirely covered by a dense laurel forest. To observe the best of this forest, turn north at the first crossroads and follow the track signposted El Cedro.

The paved track ends at the recreational area of El Cedro. The horizontal Pliocene basalts discordantly overlying the Miocene basalts at the bottom of the Barranco del Cedro form a 150 m-high waterfall that can be approached via a steep path (best in the early afternoon for photography). From this path, there is a spectacular panoramic view of the valley of Hermigua, with the Presa de Los Tiles and the trachytic needles of the Los Risquitos volcanic plug in the foreground.

Stop 7.17 Mirador de Igualero Return to the main road (TF-713). At the crossroads several kilometres farther up hill, turn left (south) towards Alajeró and at the next junction head towards Chipude. Park at the nearby Mirador de Igualero for a panoramic view to the east of the trachytic lava dome of the Fortaleza de Chipude intruding the Pliocene basalts (Fig. 7.9c). The eastern side of the dome, cut by the barranco, exposes its internal structure. The centre of the dome shows the characteristic onion-skin (radial and spherical) jointing. A short thick tongue of lava flows to the south from the dome. The view to the south offers a panorama of the radial barrancos carved on the Miocene basalts and topped with intracanyon Pliocene basalts, emitted from peripheral Pliocene vents.

Stop 7.18 Montaña del Calvario de Alajeró Return to the Alajeró–Chipude junction and take the road to Alajeró. Inside the town, take a track to the south to the foot of Montaña del Calvario. A paved path climbs to the top (810 m), from which a distant view of the islands of Gran Canaria, Tenerife and El Hierro can be seen on clear days.

The morphology of this mountain suggests an exogenous dome, formed by surface effusion of viscous lava from a central vent. La Caldera, the only

Figure 7.9 **(a)** Roque de Agando, a trachytic dome in the central plateau (stop 15); two systems of radial and concentric joints can be clearly seen. **(b)** Roque de Ojila, another characteristic and spectacular trachytic dome of the central plateau (stop 15). **(c, overleaf)** Fortaleza de Chipude (close to Chiupude village) is a trachytic dome with a thick short lava flowing towards the south (from Mirador de Igualero, stop 17). **(d, overleaf)** Roque Can, a 200 m-high trachyphonolitic vertical jointed dome located close to the town of Vallehermoso (stop 7).

189

preserved (trachytic) cinder cone of La Gomera, dated at 4.2 million years old, can be seen at the south coast (Fig. 7.10). The general view to the south and southeast is a panorama of the radial barrancos carved in the Miocene basalts, with wide flat interfluves made of Pliocene lavas in relief inversion (Figs 7.5b, 7.10). The airport is located on one of these interfluves.

The road continues to Playa Santiago and climbs to Vegaipala, to join the road to San Sebastián at stop 7.14. At Las Toscas (km-24.5) a track to the

Figure 7.10 View of the southern coast from the sea showing La Caldera, the only well preserved volcanic cone of La Gomera (dated in 4.2 million years). In the far distance to the left, the trachytic plug of the Fortaleza de Chipude.

left descends to the Barranco de Benchijigua, first on Pliocene (peripheral vents) lavas and later on the deep Miocene shield, with abundant trachytic plugs and the spectacular trachytic ring complex of Benchijigua.

Sailing around the island

Several boats in the harbour of Valle Gran Rey are available for cruises around the island. Trips can be organized during most of the year along the coast to Los Organos, in the northwest of the island (see Fig. 7.3).

The cruise to Los Organos offers views of the cliffs and the rockfall scarp at Valle Gran Rey and the coastal Miocene and Pliocene basalts at Alojera (Fig. 7.11) and, farther north, the pre-shield submarine volcanics and plutonics at Arguamul.

Figure 7.11 View from the sea of the western coast near Playa de Alojera.

191

If the weather is clear and the cruise continues around the island, there are spectacular views of the pre-shield unit until reaching the rockfall of Agulo and the mouth of the Barranco de Hermigua. Thereafter, the cliffs present a spectacular sequence of the discordant Miocene and Pliocene basalts as far as Valle Gran Rey.

Stop 19 Los Órganos North of the Playa de Arguamul the trachytic dome of Roque de Los Órganos appears (Fig. 7.12). Erosion has removed half of the dome, revealing its internal structure of regular, vertical, columnar joints, like the pipes of an organ. This jointing is the minimum-effort fracturing (three radial 120° fractures) of an isotropic mass of lava caused by thermal stresses as the lava contracts while cooling slowly and without disturbance, resulting in the formation of predominantly hexagonal, well developed columns (colonnades).

The "organs" of La Gomera are similar to those at the Devil's Postpile in California, the Giant´s Causeway on the coast of County Antrim in the north of Ireland, and Fingall's Cave on the island of Staffa, off the west coast of Scotland. Light for photographs is best in the late afternoon.

Figure 7.12   View of Los Órganos, a partially eroded trachytic dome on the northwest coast.

Chapter 8

La Palma

La Palma, the second-highest island in the Canaries, rises from the bed of the ocean 4000 m deep to 2426 m above sea level (Roque de Los Muchachos). With an area of 706 km², it is one of the steepest island volcanoes in the world, with average slopes above sea level commonly exceeding 15–20°. La Palma and El Hierro, the westernmost and youngest islands of the archipelago, are in the juvenile stage of shield building. Both islands are coeval and form a dual line of island volcanoes, similar to the Kea and Loa lines in the island of Hawaii.

La Palma, placed directly above the hotspot that has produced the Canarian archipelago, is one of the best examples of how volcanic oceanic islands can be considered as highly productive and successful seamounts that have grown above sea level. In La Palma the submarine or seamount stage of growth is very well exposed as a result of uplift and tilting of the submarine volcanic rocks. The subaerial volcano that developed on top of the seamount, the northern shield, is an excellent place in which to see the complete evolution of a shield volcano, from the uplifted and tilted submarine volcanic rocks beneath to a terminal central volcano of differentiated felsic lavas and explosive eruptions. The progression of the shield volcano towards increasing steepness and overgrowth resulted in at least two giant landslides. The last of these still dominates the morphology of the island. The volcanic history of La Palma has continued to the present with the formation in the past 120 000 years of the elongate north–south-trending Cumbre Vieja, presently the most active volcano in the Canaries.

The rock composition of La Palma volcanoes is typical of **oceanic-island basalts**, evolving from basanites and alkali basalts to tephrites, phonolites and trachytes. The greatest volumes of felsic (trachytic and phonolitic) lavas are located in the final stages of development of the northern shield, and as many domes and lava flows interspersed with basaltic rocks in the Cumbre Vieja volcano.

La Palma became a famous geological location after the work of Leopold von Buch (1825) on the genesis of the Caldera de Taburiente. Charles Lyell (1855) considered this feature to be the prototypical erosion caldera. The Caldera, as it is known, supplied Lyell with the term "caldera", to which he gave a general application and which has since mutated into a genetic term (a large crater produced by central subsidence, as in the Miocene caldera of Gran Canaria and the younger calderas of Tenerife). As discussed later, the Caldera de Taburiente was in fact initiated by a giant landslide, although it was later enlarged by retrogressive erosion.

The geomorphology of the island, shown in the shaded-relief image of Figure 8.1, is characterized by two distinct volcanoes: the Pliocene–Pleistocene northern shield and the Pleistocene–Holocene southern Cumbre Vieja volcano (Fig. 8.3), which are separated by large depressions (the Caldera de Taburiente and the Valle de Aridane, formed by the Cumbre Nueva gravitational or lateral collapse, as discussed on p. 201). This feature produces the saddle observed in the profile of La Palma when approaching the island by air or sea, also clearly visible from Tenerife.

The northern shield (2426 m) became extinct at about 400 000 years ago. The Cumbre Vieja Rift (1949 m) is, as mentioned earlier, the most active volcano in the Canaries, with the most recent eruption in 1971.

Figure 8.2 is based on studies that defined the two main geological edifices and their main stratigraphic units (Table 8.1).

Table 8.1   Volcanic stratigraphy of La Palma.

Unit		Lithologies	Age range	Locality
2. Shield-building stage	2.4 Cumbre Vieja	Basalt, basanite, trachyte phonolite	0–120 ka 1971 most recent eruption in the Canaries	Southern half of La Palma, from El Paso to Fuencaliente
	Erosion			
	2.3 Bejenado *Giant landslide*	Basanite, tephrite, phonolite	0.56–0.49 ka	Eastern wall of the Caldera de Taburiente (north of Los Llanos de Aridane)
	2.2 Taburiente	Basalt, tephrite, phonolite	1.2–0.4 Ma	
	Large-scale gravitational collapse			
	2.1 Garafía	Basalt	1.7–1.2 Ma	Barranco Los Hombres, Franceses, Gallegos, El Agua, Jieque
	Uplift—tilt—erosion			
1. Submarine –building stage (seamount)	1.2 Intrusives Subvolcanic intrusions and dykes	Basalt, trachybasalt, phonolite, trachyte (keratophire)		Lower part of the Caldera de Taburiente and Barranco de Las Angustias
	1.1 Submarine Series Basaltic pillows and volcaniclastics		Pliocene foraminifera	

Figure 8.1 Shaded relief image of La Palma showing the main geomorphological and tectonic features (image GRAFCAN).

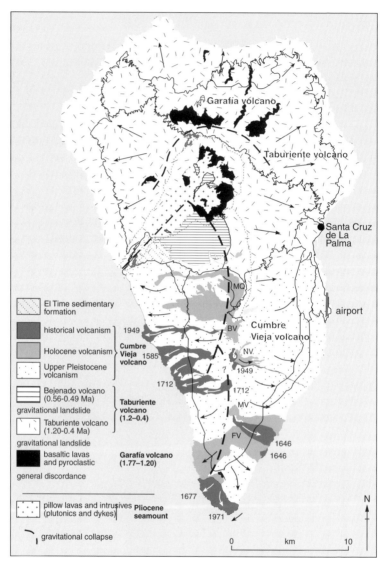

Figure 8.2 Simplified geological map of La Palma.

The following labels appear on the map:

Garafía volcano

Taburiente volcano

Santa Cruz de La Palma

airport

MQ

BV

Cumbre Vieja volcano

NV

1949

? 1949

1712

1712

MV

? FV

1646
1646

1677

1971

Legend:

El Time sedimentary formation

historical volcanism ⎤ 1949
Holocene volcanism ⎬ Cumbre Vieja 1585
Upper Pleistocene volcanism ⎦ volcano
1712

Bejenado volcano (0.56-0.49 Ma)
gravitational landslide
Taburiente volcano (1.20-0.4 Ma)
gravitational landslide

basaltic lavas and pyroclastic

general discordance

Taburiente volcano (1.2–0.4)

Garafía volcano (1.77–1.20)

pillow lavas and intrusives (plutonics and dykes)   Pliocene seamount

gravitational collapse

N

0   km   10

The northern shield of La Palma

The northern shield is composed of two clearly distinct main geological units: the submarine and the subaerial stages of growth, the latter including several successive overlapping volcanoes.

The submarine volcano

The submarine stage of growth of La Palma outcrops only inside the Caldera de Taburiente. A clearly distinct change in slope at about 1500 m above sea level marks the transition from the older and more weathered submarine volcanic rocks to the younger subaerial lavas that form the very steep cliffs that rim Taburiente (see Figs 8.1, 8.4).

The submarine volcano or seamount is formed by **pillow lavas** and hyaloclastites of basaltic composition, densely intruded (>75% of dykes in some parts) by trachytic and phonolitic dykes and **domes**, and associated subvolcanic rocks (gabbros). As mentioned before, this formation has been recognized as the seamount stage of La Palma. A complete 3.6 km-thick cross section through the seamount, from deepwater facies (>1800 m water depth at time of eruption) to the final shallow subaqueous stage, is observable in a tilted (to 50° southwest) sequence exposed along the course of the Barranco de Las Angustias. The oldest and deepest rocks are predominantly *in situ* pillow lavas with 20–50 per cent of dykes and sills, whereas in the uppermost 1100 m of the section hyaloclastites and **breccias** predominate, corresponding to the shoaling stage. The seamount displays a notable prograde metamorphic overprint, ranging from low-temperature alteration in the upper part of the section (lower course of the barranco) to **greenschist-facies metamorphism** in the lower parts of the section (green rocks at the upper course of the barranco).

The Pliocene age of the seamount has been defined by foraminifera found in hyaloclastites and dated at about 3 million years old. Endogenous growth and intrusion subsequently uplifted and tilted the seamount: hence, the deepwater stages are exposed in the upper course of the Barranco de Las Angustias. The submarine formations are topped by a thick sequence of **volcaniclastics** (breccias, agglomerates and sediments). These rocks correspond to the relatively long period of intensive erosion and explosive (**hydrovolcanic**) fragmentation in the initial emergent stage, prior to the isolation of the volcano from the sea and the formation of a resistant lava cap that allowed the island to develop. Thus, the subsequent subaerial volcanoes forming the northern shield from about 1.7 million years ago rest in erosive and angular discordance on the submarine basement, an evolution comparable to that of the seamount growth stages of Fuerteventura and La Gomera, a common pattern in the Canaries.

Figure 8.3 View from the rim of the Caldera de Taburiente (stop 8.18) of the three volcanoes that form the island of La Palma.

Figure 8.4 Oblique aerial view of Caldera de Taburiente and Bejenado volcano.

After emergence and the associated intense erosion, the subaerial part of the shield developed between about 1.7 and 0.55 million years ago, with limited activity continuing to about 0.4 million years ago. The catastrophic collapse of the southern flank of the shield at about 1.2 million years ago separates two successive volcanoes: the Garafía (pre-collapse) and the Taburiente (post-collapse). Deep erosion of the shield from about 0.6–0.5 million years ago onwards carved spectacular radial canyons (see Fig. 8.1), in which the internal structure of these volcanoes is well exposed.

The subaerial volcanoes
The Garafía volcano Completely covered by the overlying Taburiente volcano, this volcano outcrops only through erosional windows in the latter, predominantly in the natural reserve of Pinar de Garafía, from which it takes its name. The best exposures can be found at the head and walls of the deepest barrancos: barrancos de Las Grajas, Barbudo, Los Hombres and Franceses to the north; Barranco de Jieque to the southwest and Barranco del Agua to the northeast (see Figs 8.1, 8.2). The exposed sequence through the Garafía volcano forms a 400 m-thick sequence of very steeply (20–35°) and radially outward-dipping lava flows. The predominant types are thin basaltic pahoehoe flows, with profusely interbedded layers of basaltic lapilli. Dykes are abundant and follow a radial pattern. In the main outcrops, the 300–350 directions are predominant.

The limited exposure of the volcano makes the reconstruction of its original extent and height difficult. However, the structural features

199

Figure 8.5 Shaded relief image showing the Garafía and Cumbre Nueva landslides and the submarine prolongation of the Cumbre Vieja Rift (courtesy Roger Urgeles).

mentioned above, which can be also observed inside many water tunnels or **galerías**, define a conical edifice, 23 km in diameter and about 2500–3000 m high, developed on top of the underlying seamount. At about 1.2 million years ago the southern flank of the volcano collapsed into the sea. The debris-avalanche deposits that resulted have been recognized in the submarine flanks of the edifice and on the ocean floor in high-resolution side-scan sonar images (Fig. 8.5).

The Taburiente volcano The catastrophic flank failure of the Garafía volcano was not followed by a pause in the volcanic activity of the northern shield (unlike the situation in younger parts of La Palma and also in El Hierro). Instead, intense and continued activity filled the collapse embayment and spilled lavas over the northern flanks of the volcano. When activity finally ended at about 0.4 million years ago, a thick sequence of lava flows had accumulated to form the 3000 m-high (above sea level), Taburiente volcano, 25 km in diameter (see Figs 8.1, 8.2).

During the filling of the collapse embayment, flows ponded between the centre of the growing volcano and the steep collapse scarp, forming a 400 m-thick sequence of horizontal lavas. Subsequent differential erosion removed much of the pre-collapse sequence (in a fine example of inversion of relief) and left a central plateau of horizontal lavas perched in the centre of the volcano (see also Ch. 9). This geological formation hosts the main aquifer of the island; abundant springs flow from the base of the formation into the Caldera de Taburiente, feeding the permanent stream in the Barranco de Las Angustias. The Marcos and Cordero springs, located at the headwall of the Barranco del Agua at the steep contact between the plateau lavas and the underlying Garafía volcano, have the highest flow rate of all the springs in the island (310 litres per sec).

The growth of the Taburiente volcano was marked by two trends that shaped the form and structure of the island. The first, common to oceanic-island volcanoes in general, is the progressive development of rift zones. Eruptive vents, initially dispersed on the flanks of the Garafía volcano, progressively concentrated into rift zones underlain not by faults (as in tectonic rifts) but by swarms of parallel dykes. As mentioned on p. 7, these define triple or "Mercedes star" rift-zone structures, probably reflecting initial least-effort fracturing (at 120°) by magmatic doming. Once started, the rifts tend to concentrate the eruptive activity: the more dykes inject along the rifts, the easier it is for the new dykes to wedge their path parallel to the main structure. Gravitational extension stresses induced at the axes, as the rift grows higher, cause the rift zones to self-perpetuate while volcanic activity continues.

The initial evolution of the Taburiente volcano was similar to that of Tenerife and El Hierro, where the rift zones form increasingly higher and steeper "dorsales" (the local name for high volcanic ridges). The process could have ultimately shaped the northern shield into a three-pointed island volcano similar to Tenerife or El Hierro. However, the process was modified by the second trend, unique to La Palma among the western Canary Islands: southward migration of the centre of volcanic activity. This left the northern shield extinct before the rift zones could mature and so reshaped the morphology of the island into its present form.

The Cumbre Nueva rift and collapse   In the final stages of development of the Taburiente volcano, from about 0.8 to 0.7 million years ago, the southward migration of volcanism was first manifested by a concentration of eruptions in the southern or Cumbre Nueva rift zone of the volcano. The rift developed very rapidly into a north–south elongated ridge or **dorsal**, which may have reached 2500 m above sea level, as deduced from upslope projection of the lava flows at the east and west flanks of the ridge. About 560 000 years ago, the age of the youngest lavas at the top of the Cumbre Nueva ridge, the rift became unstable and its western flank collapsed into the sea (Fig. 8.5). The volume of rocks removed by the collapse was at least 180–200 km^3 (and more may have been removed below sea level), leaving an arcuate escarpment, locally known as the Dorsal de Cumbre Nueva, and an open-to-the-sea embayment, the present Valle de Aridane. The northwestern boundary of the collapse, which is now entirely eroded away, coincides with the line of the Barranco de Las Angustias.

The Bejenado volcano   This small post-collapse volcano defines the 1854 m-high southeast wall of the Caldera de Taburiente (see Figs 8.1, 8.4). However, most of the wall corresponds to the seamount, and lavas of this volcano form only the top 400 m. Although this upper section resembles the northwest wall of Taburiente, it is geologically very different. As shown in the sketch of Figure 8.6, the top of the western wall is made of lavas that erupted before the Cumbre Nueva collapse, whereas the eastern side is topped with lavas that filled the collapse embayment.

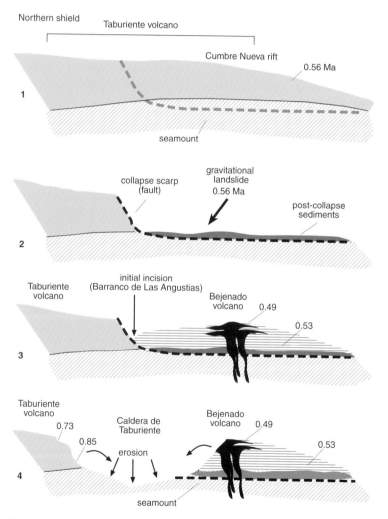

Figure 8.6   The formation of the Barranco de Las Angustias and Caldera de Taburiente. The numbers (e.g. 0.49) indicate age in millions of years.

The Bejenado lavas rest on a sedimentary formation, including the debris-avalanche deposits from the Cumbre Nueva collapse, and from rockfall and alluvial sediments, probably derived from the erosion of the unstable collapse escarpment. This formation is observable only in a small outcrop in the Barranco de Las Angustias, but has been crossed in several boreholes drilled for water research in the area of Los Llanos de Aridane.

The main part of the Bejenado volcano has been removed by the erosion that enlarged the Caldera de Taburiente. Volcanic agglomerates and explosive hydrovolcanic deposits occur at the base of the volcano, with steeply southward-dipping (20–35°) basanitic and basaltic lavas (pahoehoe and aa) forming the remaining southern flank of the volcano. Scoria beds and hydrovolcanic deposits at the top indicate proximity to eruptive vents, but these have themselves been removed by erosion. The lavas evolved to more felsic compositions (tephrites and phonolites) in the terminal eruptions, some of them from lateral vents, such as the Montaña de La Hiedra, a parasitic cone of tephritic lavas. These late lava flows are clearly visible on the flank of the volcano. Several parasitic basaltic **cinder cones** are spread around the Bejenado volcano, in the area of Los Llanos and El Paso, and others inside the Caldera de Taburiente. This waning activity of Bejenado may have continued to about 200 000 years ago.

The Caldera de Taburiente　This 15-km-long, 7-km-wide and 2-km-deep concavity, with precipitous bounding cliffs on most sides and a deeply dissected floor, is the most spectacular geomorphological feature of La Palma, but its origin has been the subject of controversy since the nineteenth century.

Age relationships in this part of La Palma provide important constraints in the origin of the Caldera de Taburiente. It deeply dissects the Taburiente and Bejenado volcanoes, and must therefore postdate both, but extends in a northeast–southwest direction parallel to the northwest wall of the Cumbre Nueva collapse structure; thus, the northwest and southeast walls of Taburiente are respectively composed of lavas prior and subsequent to the Cumbre Nueva collapse. It is therefore apparent that this caldera developed as erosion incised along the northwest lateral boundary of the Cumbre Nueva collapse embayment, forming the initial Barranco de Las Angustias and, subsequently, the Caldera de Taburiente. This interpretation may seem similar to the erosive model for the formation of the Caldera de Taburiente proposed by Lyell (1855) and many authors since, but the crucial factor is the collapse structure, followed by the growth of Bejenado. This pinned a deepening drainage system to the northwest wall of the collapse. As a result the Barranco de Las Angustias is about four times deeper (2000 m) than the deepest barrancos of the northern shield, despite the fact that the latter are older and in the northern flank of the volcano, the windward rainy side of the island.

El Time sediments   This 100–300 m-thick 8 km^2 formation partially occupies the lower course of the Barranco de Las Angustias. Lyell first described the spectacular accumulation of gravel and **conglomerates** in rhythmic depositional layers in 1855 as the "conglomerates of El Time".

The El Time sediments form the subaerial part of a **fan delta** deposited during the initial stages of downcutting of the Barranco de Las Angustias after the Cumbre Nueva collapse. These deposits are coeval with the building of the Bejenado volcano and its parasitic cones, and so the El Time formation consists of basaltic gravel and conglomerates derived from the rocks of the Bejenado Group. Submarine and plutonic clasts appear only in the upper part of the formation, recording the time when headward erosion cut through the Bejenado Group into the seamount and, as the Barranco de las Angustias deepened, into the top of the fan delta.

The Cumbre Vieja volcano

After the end of growth of the Bejenado volcano, and the entire northern shield became extinct, the island may have entered a period of volcanic repose, since no surface rocks 200 000–123 000 years old have been found. This may correspond to the period of most rapid growth of the El Golfo volcano on El Hierro, reflecting the alternation of activity between the two islands. However, volcanism in the south of La Palma may have built the submarine and core parts of the Cumbre Vieja (submarine volcanic ridge in Fig. 8.5) that are now concealed by younger lavas.

From 125 000 ago or so onwards, however, intense volcanic activity in the south of La Palma rapidly built the north–south-trending Cumbre Vieja ridge, with a maximum height of 1950 m above sea level and extremely steep flanks (average slope angles of 16–20°). With time, activity along the dominant north–south-trending volcanic rift of this volcano enlarged the island considerably, especially to the south. From about 80 000–20 000 years ago, the rate of growth of the volcano appears to have declined and marine erosion carved impressive coastal cliffs. After about 20 000 years ago, coinciding with the most recent **glacial maximum**, renewed activity produced lavas that descended the old seacliffs and built lava screes and coastal platforms (see Figs 8.1, 8.7). This activity occurred both on the north–south-trending rift zone and along subordinate northwest and northeast rift zones: the latter were notably active from 20 000 to 8000 years ago, but have been inactive since then, whereas the north–south eruptive rift now runs the entire length of the volcano.

Figure 8.7 Oblique aerial view of the Cumbre Vieja volcano. The length of the view is 30 km (courtesy S. Socorro).

It has been proposed that this structural reorganization of the Cumbre Vieja may reflect the onset of incipient flank instability of the volcano. Eruptions have occurred within the past 500 years along the crest of the volcano and in east–west fissures on the west flank of the volcano, suggesting that it is this flank that may be unstable. During the most recent eruption in the summit region of the volcano (1949), an array of west-facing normal faults ruptured the surface. The faults may indicate that a zone of fracturing is already present beneath the western flank of the volcano (although there may be other non-tectonic explanations for these fractures). Will the western flank of the Cumbre Vieja collapse, as did the Cumbre Nueva volcano? No seismicity or ground deformation has been observed in the present period of repose; even those who propose that the volcano is unstable agree that, if the collapse ever occurs, it will take place during an eruption (so, when the volcano is not erupting, the hazard is minimal) and that it will most probably not happen for many centuries (and many eruptions).

Historical eruptions
The Cumbre Vieja is the most active volcano in the Canaries and it ranks high in the active oceanic volcanoes of the world. At least six eruptions

Table 8.2 Main features of the historical eruptions of the Cumbre Vieja volcano.

Eruption	Location*	Date and duration	Years†	Composition	Mechanisms
Tahuya, Tajuya or Jedey	Western flank	19 May to 10 August 1585 (84 days)	–	Basanites, phonolites	Strombolian and block and ash (small nuée ardente)
Martín or Tigalate	Eastern flank	2 Oct. to 21 Dec. 1646 (80 days)	61	Basanites	Strombolian
Fuencaliente volcano (previously confused with the San Antonio volcano)	South end	17 Nov. 1677 to 21 Jan. 1678 (66 days)	31	Basanites	Strombolian
El Charco	Western flank	9 Oct. to 3 Dec. 1712 (56 days)	35	Basanites, tephrites	Strombolian hydromagmatic
San Juan (vents: Hoyo del Banco– Duraznero–Hoyo Negro)	Summit and western flank	24 June to 31 July 1949 (38 days)	237	Basanites, tephrites	Strombolian hydromagmatic
Teneguía volcano	South end	26 Oct. to 19 Nov. 1971 (25 days)	22	Basanites, tephrites	Strombolian

* In relation to the Cumbre Vieja rift.
† From previous eruption.

have occurred since the Spanish colonization of the island at the end of the fifteenth century (Table 8.2). Most of these eruptions were of short duration (<3 months), basanitic-tephritic composition and Strombolian eruptive mechanisms. Only the eruption of 1585, with juvenile phonolites, involved more dangerous **block-and-ash** events, similar to small **nuée ardente** eruptions. The Hoyo Negro summit vent of the 1949 eruption involved violent hydrovolcanic explosive phases.

A statistical prediction of when the next eruption will occur is unfeasible. Although most of the inter-eruptive periods are 22–61 years, the 1949 eruption was preceded by a very long period (237 years), in which the population lost all awareness of living in an active volcanic environment.

Logistics on La Palma

The few beaches in La Palma are small, and the sand (black basalt) is less attractive than the white or golden beaches in the older and warmer eastern islands, where the sand developed from coral or light-coloured felsic rocks. Consequently, tourism in La Palma (135 000 visitors annually of 11 million in the Canaries) is mostly aimed at nature and mountain trekking. The population of the island is relatively small (80 000 inhabitants) and mostly concentrated near Santa Cruz de La Palma and Los Llanos de Aridane. In the remainder of the island, particularly the northern shield, there are only some small towns, scattered groups of houses and isolated farms.

Consequently, package offers to La Palma are not so easy to find, since there are only a few holiday complexes and hotels in the southwest of the island (Los Llanos de Aridane, Puerto Naos) and in the Santa Cruz de La Palma area. However, most small towns have cheap and convenient apartments and pensiones. If you hire a car and are prepared to do some walking, the casas rurales are a very good choice (www.islabonita.com), available throughout the entire island and many of them close to the areas of greatest geological interest. Book well in advance.

Several scheduled flights are available daily from Tenerife or Gran Canaria, as well as a cheaper ferry from Santa Cruz de Tenerife and Los Cristianos (south Tenerife).

The island is not well served by public transport, and taxis are expensive. Hire cars are the best choice, being cheap and plentiful. A word of warning: roads are narrow, steep and winding, although not busy. You will be tempted by the many dirt tracks that wind through the forests and inside the barrancos (to serve the galerías). To that end you should hire a four-wheel-drive car and have some off-road experience. There are several very spectacular itineraries that involve descending 1000–2000 m. Retracing your steps is not feasible in these cases, but you can plan in advance for a friend or a taxi to pick you up and take you to your car.

The weather in La Palma is seldom a problem. An almost constant trade-winds regime makes the climate very mild throughout the year. In normal conditions a mar de nubes ("sea of clouds") can attach to the north and northeast flanks of the northern shield below 1000–1500 m. Climbing higher offers a magnificent and very clear view of La Palma and the neighbouring islands Tenerife, El Hierro and La Gomera. If you are unfortunate (more probable in wintertime), a western (Atlantic) storm can hit the island, with heavy rains, snow and ice in the highlands, and strong westerly winds. The opposite is African weather, known as tiempo sur ("south weather"), which is extremely hot, dry and dusty with very low visibility.

Watch out for rockfalls at scarps and barrancos – particularly after heavy rains – and keep away from the very unstable edges of the coastal cliffs. Take precautions against excess exposure to the sunshine and against dehydration. A cellular phone, with satellite coverage for the emergency number 112, can be a safety bonus.

Almost 40 per cent of the island is protected by law. The most stringent restrictions apply to the Caldera de Taburiente national park. Other protected areas, the natural parks, have fewer restrictions. However, if you plan to collect samples, ask for information and apply for a permit in advance from the Department of the Environment of the Canarian government (www.gobcan.es/medioambiente) to avoid embarrassing situations in the field or checking out at the airport.

Very precise topographic maps (1:25 000 and 1:5000) and aerial photos, showing every road and track, are available from the Instituto Geográfico Nacional (www.cnig.ign.es) and the Canarian government.[*] Geological 1:25 000 maps of La Palma by the Instituto Geologico Nacional are in press.[†]

Itineraries in the submarine volcano

As discussed in the introduction to this chapter, the interior of the Caldera de Taburiente exposes a tilted submarine volcano or seamount, incised by the Barranco de Las Angustias. Therefore, as we progress inside the caldera we go deeper into the seamount. The first major section of the itinerary includes the shallow-water and deepwater facies (1a), whereas the section farther inside the caldera comprises the plutonic core (1b).

A walk on the flank of the seamount

Start at the bus station in Los Llanos de Aridane (Fig. 8.8), where the Bejenado volcano can be seen in the background, and enter the Barranco de Las Angustias using the road signposted to the Caldera and the track down to the course of the barranco. Leaving the car in the parking lot offers two choices: turning off down course will lead into the shallowest water facies of the seamount, whereas ascending the barranco will progress towards the deepwater parts of the seamount. This itinerary follows the part of the barranco outside the Parque Nacional; there are no particular restrictions.

Walk about 1.5 km down to the Casas de La Viña, where the barranco narrows and the itinerary begins.

Stop 8.1 Casas de La Viña to the parking lot The south wall is composed of Bejenado lavas discordant on top of the seamount. Where these lavas meet the bed of the barranco there is a small outcrop of debris-avalanche deposits of the Cumbre Nueva collapse beneath them. The northern wall of the barranco is carved in the thick sedimentary formation of El Time. Some intracanyon lavas from peripheral vents of the Bejenado volcano are interbedded with the sediments; these once formed marine lava deltas. Up stream from this point to the parking lot, the bed cuts into much older rocks: submarine volcaniclastics (mostly predominant) and lavas of the seamount. Spectacular pillow lavas (Fig. 8.9) display a variety of forms and sizes. In some outcrops, different types of corals appear intercalated

---

[*] www.grafcan.rcanaria.es

[†] Printable digital colour maps of La Palma and El Hierro (up to 1:25000) can be downloaded from http://www.ipna.csic.es

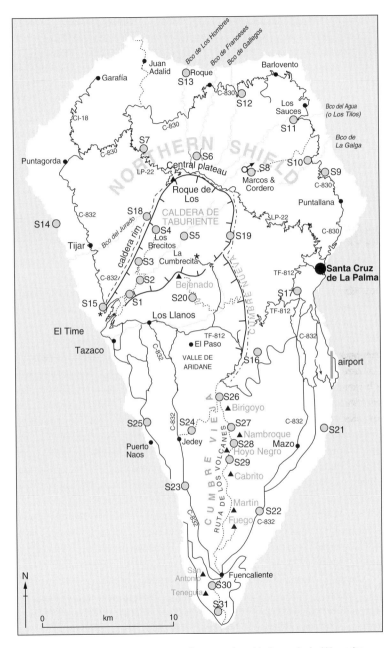

Figure 8.8 The main roads and other features referred in the geological itineraries.

Figure 8.9 Pillow lavas and dykes of the seamount inside the Caldera de Taburiente (stop 8.2).

between the pillows. These rocks are characteristic of the *in situ* depositional facies at the flanks of a seamount in shallow (shoaling) water.

Stop 8.2 Parking lot to Dos Aguas From the parking lot up stream, the trek follows a well maintained path along a gravel stream bed. The walk progresses towards the earlier stages of development of the seamount, when it was a deep abyssal hill on the ocean floor. Spectacular lava pillows with characteristic tubular and amœboidal features can be observed in the bed and in large boulders. Dykes become increasingly frequent up stream, including earlier feeders of the seamount and later (fresher) dykes that fed more recent subaerial eruptions. The dykes take up to more than 50 per cent of the rock, forming a very dense dyke swarm (see Fig. 8.9b). The white lava screens preserved between the dykes are brecciated or massive **keratophyre**.

Hydrothermal alteration increases in grade up stream at a metamorphic gradient of about 200°C per km, changing the primary components of the original trachytes and phonolites to metamorphic minerals (chlorite, epidote), which produce the characteristic green colour of the rocks.

The sheeted dyke swarm increases in density up stream, grading into the plutonic core at the centre of the caldera. The itinerary ends before that stage, at the place known as Dos Aguas ("two streams"), where the national park begins. Near the rock wall damming the stream at Dos Aguas, CO_2-rich hydrothermal water springs through fissures in the reddish hydrothermally altered rocks in the bed of the river.

The plutonic core at the centre of the Caldera de Taburiente
This itinerary starts in the same parking lot and follows the road to Los Brecitos viewpoint and car-park. A signposted path continues into the national park to the camping area, at the geometrical centre of the caldera, where several hiking trails begin. It is very dangerous to walk off the sign-posted trails; this should never be attempted without a professional guide and a permit from the national park (www.mma.es/parquesnacionales).

Stop 8.3 Hacienda del Cura–Los Brecitos viewpoint and car-park   The first stretch comprises the ascent by a narrow and winding track from the Barranco de Las Angustias to the base of the subaerial lavas discordantly overlying the seamount, at 1000 m above sea level. The road crosses the seamount submarine lavas and volcaniclastics injected by a dense dyke swarm. Rockfall and scree deposits are abundant. Past the Hacienda del Cura, the road runs close to the discordant contact with the overlying subaerial volcanic formations. Layers of basaltic pyroclasts (remains of interbedded cinder cones), agglomerates and lavas built a 500–1000 m-high vertical cliff from about 0.83 million years ago (the age of a basaltic lava at the base of the cliff near Hacienda del Cura) to 0.52 million years ago (the age of a phonolite lava at the top of the cliff at Morro de Pinos Gachos). The abundant volcaniclastic rocks, with interbedded hydrovolcanic deposits (the distinct yellow layers at the base of the cliff) show the highly explosive character of the initial stages of subaerial activity. The very dense dyke swarm of the seamount is cut at the discordant contact. The fewer dykes at the cliff are feeders of the subaerial volcanics, some of which cross the entire formation.

The best views of the Bejenado volcano overlying the seamount can be seen between the Hacienda del Cura and Los Brecitos. The afternoon is best for photographs. Note a very thick light-coloured (probably phonolitic) sill that cut the Bejenado lavas to feed the late felsic eruptions of the volcano.

Stop 8.4 Los Brecitos-Camping site   After Los Brecitos car-park (in high season parking may be difficult) the trail follows the base of the cliff for a short distance, over impressive rockfall and scree deposits, some of them very recent. These are produced by the rapid retreat of the escarpment that is progressively enlarging the Caldera de Taburiente and they frequently change the course of streams and trails, especially after heavy rains. One of the rockfalls, at Tenerra, is particularly large and has been terraced for a former tobacco plantation, now the only farm inside the national park. The trail descends for 350 m (5.5 km) to the camping site crossing a thick bush and pine forest. It is a pleasant walk, but the geological formations are exposed only in the bed and walls of the streams. These outcrops show the sheeted dyke swarm and screens of plutonic rocks, mainly gabbros (white rocks), in different stages of hydrothermal alteration.

Stop 8.5 The camping site This is located at the margin of the barranco Río Taburiente (the continuation up stream of the Barranco de Las Angustias), where the river bed opens into a spacious (200 m wide, 1000 m long) alluvial flat of sand and gravel (the Playa de Taburiente). To the east of the camping site rests a prominent group of spurs (Roque Salvaje, Roque de la Brevera Macha, Roque Capadero and Roque de las Piteras). These are 100–250 m-high blocks of volcanic agglomerates with interbedded lavas and pyroclasts. The main alignment of blocks rests discordantly on the seamount along the interfluve between the barrancos Río Taburiente and Río Almendro Amargo. They are remains of volcanics from eroded parasitic vents of the Bejenado Group filling the bed of former barrancos. The filling of the older ravine probably forced a divide and left the remnants of the fill isolated by inversion of relief.

Itineraries in the northern shield

The Garafía volcano: Pinar de Garafía

Although the Garafía volcano forms small outcrops in the bed of several deep barrancos (Jieque, La Herradura, etc.), these are difficult to access. The more accessible extensive and characteristic outcrops are in the natural reserve of the Pinar de Garafía (between the barrancos of Las Grajas and Gallegos) and in the Barranco del Agua.

Take the north road (C-830) out of Santa Cruz de La Palma. After crossing the main barranco above the town, take the very winding but spectacular road to the astrophysical observatory (LP-22), and drive 30 km up hill from sea level to 2300 m. The ascent will constantly cross steep basaltic lava flows and several dykes of the Taburiente volcano and basaltic cones (Montaña Tagoja, Montaña Llano de Las Vacas, Monte Santo, etc.) of the Puntallana–Barlovento rift zone.

A track at km-22 climbs easily to Pico de La Nieve, at the rim of the caldera (see stop 8.19).

Stop 8.6 The Tamagantera trail On arriving at the casa forestal (km-30) enter about 100 m on a dirt track to the north and park the car. Look for a stone-paved trail (the Tamagantera camino real). The trail descends over 350 m of horizontal lavas of the central plateau (Taburiente volcano) to the discordant contact with the Garafía volcano (Fig. 8.10a). The scenery on this walk is most impressive, but the trail at times can be a little slippery. Several dykes cross the entire formation to feed emission vents at the top of the volcano.

The first lavas encountered are flows of the Taburiente volcano, which spilled over the northern flank of the Garafía volcano after filling the collapse embayment. However, most of the plateau lavas were produced in a very short period after the Garafía collapse, 1.08–1.02 million years ago. The contact with the Garafía volcano is marked by a baked red layer (**almagre**) and a few dykes truncated at the contact. To the west of the trail, the heads of the barrancos de Los Hombres, Barbudo, El Cedro and Las Grajas can be observed (best with binoculars and before noon for photographs), forming an erosional window through the Taburiente volcano into the older Garafía volcano, dated here at 1.52–1.49 million years ago. The discordance is clearly marked by the differences in age and density of dykes, and by the angular contrast between the horizontal post-collapse lavas and the steep Garafía flows.

The trail continues along the Lomo de Los Corrales, crossing Garafía lavas. Within 1 km it again encounters the lavas of the flank of the Taburiente volcano, as the erosional window closes. You have the choice now of returning to the car or descending to Roque del Faro, less than 6 km distant but 800 m below. This is a long and tiring walk with a happy ending at the bar at Roque del Faro.

Stop 8.7 Sharp corner at km-39 of LP-22   Descending from the astrophysical observatory towards the town of Garafía, there is an excellent viewpoint from which to observe the Garafía volcano and the plateau lavas at the head of Barranco de Las Grajas (Fig. 8.10b), best in the afternoon for visibility and photographs.

The Garafía volcano: headwall of Barranco del Agua
This itinerary begins at El Tanque, on the northern road (C-830, near km-25) where a track signposted Manantiales de Marcos y Cordero starts.

A winding and frequently busy track climbs to the Casa del Monte (1300 m), where the trail to the Marcos and Cordero springs begins. The track climbs over the young lavas of the flank of the Taburiente volcano (0.7–0.5 million years old). However, the high rainfall on this side of the shield, almost constantly under the trade-winds regime, has favoured the deep alteration of the lavas to a clayey grey paste. Close examination reveals "ghosts" of columnar jointing and spheroidal (onion-skin) weathering features of the original flows.

Figure 8.10 Two different views of the central plateau of horizontal lavas and the Garafía volcano. **(a)** The Tamagantera trail and horizontal lavas (stop 8.6) from the LP-22 road; the arrow shows the discordance. **(b)** The central plateau hanging on the Garafía volcano lavas at the head of Barranco de Las Grajas, as seen from stop 8.7.

Stop 8.8 Trail to the Marcos and Cordero springs   This is a spectacular walk into the amphitheatre valleyhead of one of the oldest and deepest barrancos of the northern shield. The trail will enter several tunnels (a flashlight and helmet are useful) and small waterfalls (a raincoat and spare footwear make the walk more comfortable).

The trail follows the water conduit from the springs and provides a magnificent profile crossing the lava flows of the Taburiente volcano to the discordant contact and into the Garafía volcano. The first lavas to cross are relatively young flows from the top of the Taburiente volcano (0.7–0.8

million years ago), with very scant interbedded palaeo-soils and dykes. A prominent rockfall scarp has shaped the first stretch of the trail.

The path crosses 200–300 m of Taburiente volcano lavas, getting deeper into the volcano, as shown by the increasing alteration of the flows and the number of dykes. Most of the lower part of the sequence corresponds to the horizontal lavas of the central plateau.

The contact between the Taburiente and Garafía volcanoes is marked by reddish brecciated conglomerate. The thick sequence of Taburiente volcano lavas stores snow and rainwater, and the altered and less permeable sediments and volcaniclastics at the contact form a barrier that creates the most important aquifer system in La Palma. The water is forced to flow up the contact and forms the Marcos (1360 m) and Cordero (1420 m) springs, exploited since the sixteenth century. A very similar feature – a collapse embayment filled with lavas – also provides the only important aquifer of El Hierro.

The best point to observe these features is at the Cordero spring. The Garafía volcano lavas underlying the contact have been dated at 1.38 million years old below the contact, and at 1.52 million years old lower down the stream bed, corresponding to older sequences of the volcano. They are the characteristic thin and steep (20–25° northeast) pahoehoe flows, with some thicker aa flows interbedded. Dykes are abundant and follow a predominant trend parallel to the northeast (Barlovento) rift zone of the northern shield.

The Taburiente volcano: the lower flanks

Two itineraries are suitable for observing the main features of the Taburiente volcano. The first circles the volcano using the roads C-830 from Santa Cruz de La Palma to Puntagorda (see Fig. 8.8), the C-832 to Los Llanos de Aridane and the TF-812 to end the round trip in Santa Cruz. The second itinerary is a long and impressive walk following the rim of the Caldera de Taburiente and the Cumbre Nueva collapse escarpment.

This long drive, which may take more than one day, encircles the base of the Taburiente volcano, crossing the main rifts and the many deep canyons incising the volcano in a radial network.

Stop 8.9 Barranco del Agua Leaving Santa Cruz de La Palma towards the north, the road passes to the right of the basaltic cone of Montaña Tenagua, a vent of the Puntallana rift zone. The flows of the late vents of the rift overflowed the cliff and formed a flat coastal platform of lavas (punta llana, "flat point").

At Barranco del Agua (another barranco of the same name is discussed at stop 8.11), the first deep barranco after the town of Puntallana, a sequence of valley-filling lavas appear hanging on the canyon walls as a result of differential erosion of the former barranco. The lower formation has cinder cones and dykes, whereas the upper one is made of lava flows. This feature is characteristic of the Taburiente volcano (as discussed in the geological introduction) and it repeats in most of the canyons around the volcano.

You may see holes drilled in the rock. These have been made to measure the **geomagnetic polarity** of the lavas to define and accurately date the **Matuyama/Brunhes** (M/B) boundary (0.78 million years old). You will also see such boreholes in many of the canyons, marking the change in the polarity of the geomagnetic field and the epoch in which the volcano reorganized into rifts.

Stop 8.10 The La Galga hydrovolcanic tuff ring and deposits Continuing north, and after crossing many cinder cones of the Puntallana rift zone, the road arrives at the small village of San Bartolomé, with the cinder cone of La Galga to the east. Attached to the south base of the cone, the remains of a **lahar** deposit is visible at outcrop. This is part of a broad explosive and laharic formation that originated from a hydrovolcanic vent, a 1.5 km-wide and deeply eroded crater, located up hill (see also pp. 77–80 for details on tuff rings and tuff cones).

Entering the (next) Barranco de La Galga, another remnant of the deposit forms a pinnacle at the top of the interfluve, a clear case of inverted relief of a former barranco channelling the laharic flows. To reach the remnants of the eruptive centre, follow the track at the bottom of the barranco westwards to the Cubo de La Galga (about 2 km drive or walk). Part of the vent outcrops at a gravel quarry, with well preserved **surge deposits**, lahars and **ignimbrites**.

These deposits must have completely disrupted the drainage system, channelling along the barrancos. They have subsequently been left isolated between the later barrancos as the relief inverted.

Stop 8.11 Barranco del Agua or Los Tiles This barranco (another barranco of the same name is discussed at stop 8.9) and the following ones at stops 8.12 and 8.13 show spectacular deep incisions in the entire flank of the Taburiente volcano. They are interesting because of the geology and the magnificent scenery.

The upper and lower formations of the volcano intersect at the bottom of road C-830 (km-25), show the recent palaeomagnetic studies carried out in the Matuyama/Brunhes boundary, the most recent main transition from reverse to normal polarity of the Earth's magnetic field at about 0.78 million years ago. At the bend of the road, another tarmac road running westwards follows the barranco to the picnic area and into the El Canal y Los Tiles (a UNESCO Reserve of the Biosphere since 1983), a thick forest with many endemic species of flora and fauna. There is an information centre for the reserve nearby.

The lower walls of the barranco intersect fossilized cinder cones and dykes of the early Taburiente volcano. From the recreation site, a trail follows the course of the barranco into a spectacular narrow and deep gorge, carved into the underlying Garafía volcano; the dykes are densely packed and the lavas are 1.44 million years old.

Stop 8.12 Barranco Gallegos   After leaving behind the group of cinder cones of the Barlovento rift zone and town, the road enters the Barranco Gallegos (Fig. 8.11), a deep canyon similar to the previous one. In the wall, the road intersects the entire Taburiente volcano (lavas dated at 0.56–0.88 million years ago). The Garafía volcano, again densely packed with dykes, outcrops at the bottom of the canyon, where lavas are 1.20–1.44 million years old.

Stop 8.13 Barranco de Los Hombres   This barranco, where the scenery is spectacular, provides the most complete section into the subaerial stage of the northern shield. The road now crosses the Barranco Franceses (similar geological setting, with the Garafía volcano outcropping at the bed of the barranco). The scar of a recent rockfall caused by construction work on the tunnel is visible on the eastern wall.

As you leave the barranco, a track descends to the village of Franceses and from there to the Barranco de Los Hombres. The geological sequences are the same, although more complete since, after filling the Garafía collapse embayment, the lavas reached their spill point and accumulated here. These lavas, dated 948 000–833 000 years old and corresponding to the **Jaramillo event**, may represent the first lavas to overtop the collapse scar left by the Garafía collapse and exit the central plateau through former barrancos on the flank of the Garafía volcano, particularly via the early Barranco de Los Hombres.

The road ends at the coast in a wide embayment (La Fajana) originated by a large rockfall that involved about $0.05\,km^3$ of bedrock. Rockfalls are a very frequent and efficient mechanism of erosion and retreat of the coastline. The most spectacular example is the Playa de La Veta rockfall described in the next stop.

Stop 8.14 Playa de La Veta   The falling of newly detached segments of bedrock from cliffs considerably accelerates retreat of the coastline or inland escarpments in oceanic-island volcanoes. There are many such

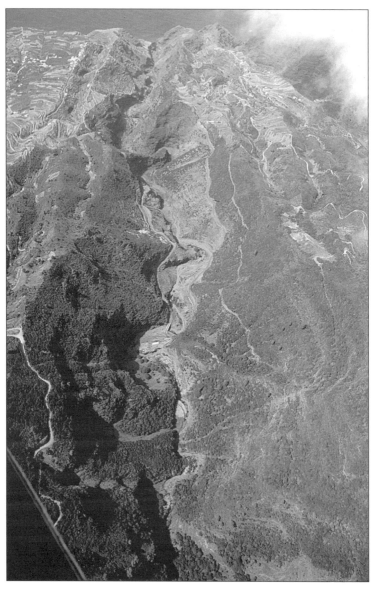

Figure 8.11 Oblique aerial view of the Barranco Gallegos (stop 8.12). The stretch of the barranco in the photo is 6.5 km.

features along the coast of the northern shield of La Palma and the seaward coast of La Gomera. Similar rockfalls of spectacular dimensions can be seen in the Hawaiian Islands (e.g. the Napali coast in Kauai or the northern coast of Molokai) and in the walls of the cirques of Piton des Neiges volcano in Réunion Island.

One of the best examples in the Canaries is at the Playa de La Veta (Fig. 8.12). Drive round the northwest flank of the shield, where you will find a succession of barrancos similar to those described (although not deep enough at this side to cut into the Garafía volcano). After crossing the northwest-trending Garafía–Puntagorda rift zones and towns, take the C-832 road towards Los Llanos de Aridane. Pass the km-72 post and drive into a very steep winding track (opposite Bar La Guagua) that runs down to the main north–south water duct and track. A few hundred metres to the south the track meets a cement path that plunges down the cliff (there is a car-park farther down, but you may prefer to walk; take plenty of drinking water). The track changes into a narrow path down the cliff and over two massive rockfalls. Most of the original bedrock features (lavas, dykes), as well as the scar at the wall, are still clearly discernible.

There is a beautiful small beach, developed between the collapsed blocks, where you can swim and rest before the return climb. The best photographic conditions are before noon.

Figure 8.12 Rockfalls of Playa de La Veta, at stop 8.14.

Stop 8.15 The Mirador de El Time   Back on road C-832, drive through the town of Tijarafe (El Pueblo on some maps) and the Barranco del Jurado to the El Time viewpoint ("time", from the Berber "timme", "edge"). On clear days in the late afternoon there is a magnificent panorama of the Caldera de Taburiente and Bejenado volcano to the right and the Valle de Aridane, with several peripheral cones of the Bejenado Group and a lineation of three cinder cones of the Cumbre Vieja volcano directly in the foreground.

Looking down to the Barranco de Las Angustias river bed, the sedimentary formation of El Time can be observed attached to the walls of the canyon. Alternating terraces in the delta plain at both sides of the barranco were originated by the changing course of the river as headward erosion forced the barranco to cut through the already inactive fan delta.

Interesting examples of the complex sedimentary facies characteristic of a former deltaic environment are observable as the road descends to the bottom of the canyon. Before crossing the bridge you can turn to the right and follow the road to the Puerto (harbour) of Tazacorte. The attachment of the sediments to the Taburiente volcano lavas forming the wall of the canyon is clearly visible at the west wall. At the other side of the river bed and close to the beach, the base of the sedimentary sequence comprises fine-grain layers (silt, clays) with traces of subaqueous turbidity flow and bioturbation, suggesting deposition in a pro-delta environment.

You can walk the lower course of the Barranco de Las Angustias up stream following a trail that begins just below the bridge on C-832. Intra-canyon lava flows from parasitic vents of the Bejenado Group appear interbedded in the sediments. The walk ends at stop 8.1 (Casas de la Viña).

Stop 8.16 The road tunnel at Cumbre Nueva   Continuing through the towns of Los Llanos de Aridane and El Paso towards the tunnel through the Cumbre Nueva ridge, the road passes over a plain of Cumbre Vieja lavas, cinder cones and sediments derived from the Cumbre Nueva and Bejenado, which have infilled the Cumbre Nueva collapse embayment.

Just as the road begins to climb up the escarpment towards the Cumbre Nueva tunnel outcrop, there are east-dipping lavas (830 000 years old) and densely packed north–south-trending dykes corresponding to the core of the collapsed Cumbre Nueva rift zone. At the other end of the tunnel – there is usually a dramatic change in weather conditions from the dry, sunny western side into thick clouds, the same clouds that can often be observed cascading from the top of the Cumbre Nueva ridge – the road passes over eastward-dipping lavas, corresponding to the flank of the Cumbre Nueva rift. These lavas clearly flowed on very steep slopes, when the high and unstable volcano was close to collapse (the top lavas are 0.56 million years old).

Stop 8.17 La Caldereta Continuing towards Santa Cruz de La Palma you will see below the wide crater of the La Caldereta tuff ring. Drive to the Mirador de La Concepción viewpoint at the top of the cone. Looking up hill to the west to catch a panorama of the Cumbre Nueva ridge, you can observe lavas of the Taburiente volcano that flowed down slope to pond against and encircle La Caldereta. From the mirador you have a spectacular view inside the 1300 m-wide, 400 m-deep crater. This tuff ring is very similar to the La Galga volcano described in stop 8.10, but it is better preserved, being protected from erosion by the encircling lava flows.

The cone is mainly hyaloclastites (characteristically yellow), with a wide display of explosive hydrovolcanic features (ballistic block sags, cross bedding, palagonitized hyaloclastites, etc.), visible as the road enters the harbour (in the cliff crossed by the road tunnel). From the mirador you can observe a small nested Strombolian cone and also a short lava flow ponded in the crater. This clearly shows that the eruptive conduit became isolated from the sea water in the late stages of the explosive eruption, which changed to a "normal" Strombolian eruptive mechanism as the supply of water diminished (best for photography before noon).

The rim of the Caldera de Taburiente (late stages of the Taburiente volcano) There are two full-day walk routes, both of which require some planning. Do not attempt either route in doubtful weather conditions. Both walks, along well maintained trails, start from the same point at the Roque de Los Muchachos information cabin (a good precaution is to report your itinerary). You will need plenty of drinking water and sun protection (hat, sunscreen, etc.) but it can be cold as well. Start early.

The Roque de Los Muchachos divides the itinerary into two walks: west, ending at the El Mirador de El Time (stop 8.15), and east, ending at the Ermita Virgen del Pino, near El Paso. The west walk (down slope) is to observe the interior and eastern walls of the caldera and the northern part of the Bejenado volcano (best for photography from noon on clear days). The east walk (mostly down slope with some climbs) is to observe the interior and western walls of the caldera (Fig. 8.13) and the Bejenado and Cumbre Nueva volcanoes (best for photography in the morning).

Stop 8.18 Roque de Los Muchachos–Mirador de El Time Not strictly a "stop" but a 16 km walk on the rim of the caldera, descending from 2420 m to 500 m. The best plan is to have someone take you here early in the morning and pick you up in the Mirador de El Time in the late afternoon.

The first part of the trail runs among cinder cones of the upper part of the Garafía–Puntagorda rift zone and some of the highest elevations of the caldera rim (Roque Chico 2772 m, Roque Palmero 2310 m, Morro Pinos

Figure 8.13 Panoramic view of the northern and western walls of the Caldera de Taburiente from the Roque de Los Muchachos, stops 8.18 and 8.19.

Gachos 2180 m and Somada Alta 1926 m). These elevations are formed by the youngest felsic lavas of the Taburiente volcano (the phonolites at Morro Pinos Gachos have been dated at 0.52 million years old). Near Somada Alta are block and ash deposits similar to those at Jedey (stop 8.24).

The walk continues down hill to the Mirador de El Time (stop 8.15) on the youngest lava flows of the western flank of the Cumbre Nueva rift.

Stop 8.19 Roque de Los Muchachos–Ermita Virgen del Pino   This is another 23 km walk on the rim of the caldera, which descends from 2420 m to 900 m. You can divide the walk into two parts – two different days – using the track to Pico de La Nieve (see stop 8.6).

From Roque de Los Muchachos to Pico de La Nieve the walk follows the caldera rim and the uppermost lavas of the Taburiente volcano, with lavas alternating with interbedded cinder cones and abundant dykes. The lavas are mainly basalts and tephrites, but some of the top flows are phonolites. Cinder cones and lava flows crossed by dykes form the visible part of the wall. Down at the footwall a change in slope reveals the older seamount. From many points a spectacular panoramic view can be had of

the caldera and, at the bottom, the Taburiente River with the alluvial flat of Playa de Taburiente (stop 8.5).

From Pico de La Nieve to Pico del Cedro the same geological formations outcrop. From Pico del Cedro, with excellent views of the Caldera and of the Bejenado volcano, the walk follows the Cumbre Nueva collapse escarpment facing the collapse embayment and the Cumbre Vieja volcano (one of the best places to photograph the Cumbre Nueva and the northern part of the Cumbre Vieja Rift). In the distance can be seen the pre-historical (1480) Montaña Quemada and Birigoyo volcanoes. As the slope flattens, a trail turning right and down slope descends the escarpment to the Ermita ("chapel") Virgen del Pino. The lavas of the escarpment, forming the flank of the eastern flank of the Cumbre Nueva rift, are all of basaltic composition and erupted from 0.6 to 0.56 million years ago.

The last stretch of the trail runs on scree deposits attached to the cliff.

The Bejenado volcano

Stop 8.20 The climb to the top of the Bejenado volcano   The itinerary begins at the national-park visitors centre, east of El Paso. Follow the road marked La Cumbrecita and, at the crossroads, the track signposted Valencia.

223

On the right-hand side of the road several parasitic cinder cones of the Bejenado Group can be seen, deeply eroded and partially covered by lava flows. As the road begins to climb the southern flank of the Bejenado volcano, it passes the basaltic cinder cone of La Montañita on the right-hand side, and the Montaña de la Hiedra (a cinder cone with tephritic lavas with hauyne crystals) to the left. The road, on fresh, steep pahoehoe and aa lavas of the Bejenado, turns left to a stretch of horizontal track. About 1 km ahead, a trail begins the ascent to the top of the Bejenado volcano. On arriving at the first open view of the caldera (at Los Rodeos), turn right to proceed to the Pico de Los Cuervos, a part of the terminal felsic (phonolitic) eruptive centres of the Bejenado.

Return to Los Rodeos and follow the trail to the west where the final climb to the top of the volcano begins. The trail continues up over pyroclastic beds, some of them hydrovolcanic, and some dykes. On arriving at the top, on very steep (30–40° dipping) felsic lavas, the view is magnificent: on a clear day this is one of the best possible panoramas of the caldera. The view to the south is also one of the best sights of the Cumbre Nueva collapse embayment and the Cumbre Vieja volcano.

The Cumbre Vieja volcano

Two itineraries (two days) are described to observe, respectively, the outer flanks of the Cumbre Vieja volcano (predominantly lava flows and coastal hydrovolcanic vents) and the summit crest, composed of many perfectly preserved eruptive vents. The first itinerary follows the C-832 road from Santa Cruz de La Palma (or a parallel road closer to the coast signed to San Antonio village and the archaeological site of Cueva de Belmaco, joining the C-832 past the village of Tigalate), to Fuencaliente and Los Llanos de Aridane. This excursion will consider the stratigraphy of the Cumbre Vieja volcano, defined using changes of sea level exposed in the coastal cliffs. The second itinerary follows the trail locally known as the Ruta de Los Volcanes, along the Cumbre Vieja volcano ridge, to observe the main vents of the historical and geologically recent eruptions of the Cumbre Vieja volcano.

The lower flanks of the Cumbre Vieja Rift: driving around an active volcanic rift *Stop 8.21 The Montaña Goteras hydrovolcanic eruption and Las Salineras coastal platform* The itinerary runs south towards the village of San Simon, where a track leads down slope to Montaña Goteras, a tuff cone with magnificent features of hydrovolcanic explosive eruptions (see stop 8.17). Splendid photographs can be obtained, particularly in the morning.

Follow the track to Punta Las Salineras (often signed as Salamera or Salemera), which runs on top of the coastal platform of young lavas built up since the end of the most recent glaciation. Scree-forming lava flows fossilize the cliff carved before and during the most recent glacial maximum. The palaeo-cliff, observable through windows in the young lava screes, can be inferred by the change in slope at the back of the platform.

Stop 8.22 The 1646 eruption (the pre-Martín and Martín volcanoes) Drive on to the village of Monte de Luna (C-832, km-22.5); previously, at km-9, the road crosses the eastward lava flow from the 1949 eruption (the Duraznero eruptive vent), which stopped just 50 m from the shoreline.

In Monte de Luna the road enters a very recent lava flow, sparsely covered by vegetation and originating in a group of cinder cones higher up slope. Lava channels and levees are very distinct in these flows, which, along with the upper cinder cones, had been thought to belong to the eruption of 1646. The true 1646 flows are crossed at km-23. At the boundary you will note important changes, such as the absence of vegetation in the latter. However, the most important evidence for two distinct eruptions is the presence of pre-Hispanic archaeological remains in lava tubes north of this boundary. A splendid, albeit small, eruptive vent of the 1646 eruption (El Búcaro) lies just below at the coastline. It can be approached by a very steep dirt track from Las Caletas, near Fuencaliente (not suitable for cars).

The 1646 lava flows end sharply at km-24, where the road enters the lavas of the Fuego volcano, a prehistoric (1100 BC) eruption from vents located higher (1200 m) up hill. These lavas (tephrites) formed one of the largest single eruptions of the Cumbre Vieja volcano. They originate from a complex of vents on the main rift zone, which is located on top of an older phonolitic dome (Roque del Pino de La Virgen), an example of the relationship frequently seen in the Cumbre Vieja volcano between fractured phonolitic domes and recent eruptions (see stop 8.24 for details). Flows descended from these vents to both the eastern and western coasts of the Cumbre Vieja. A thick phonolitic lava from the Roque outcrops at the roadcut at km-25.

Stop 8.23 Views of the west coast of Cumbre Vieja volcano Past Fuencaliente the road continues along the leeward western flank of Cumbre Vieja. A stopping point is the Mirador de El Charco, from which the west coast of La Palma can be seen, as far north as the El Time cliff at the mouth of the Barranco de Las Angustias. The mirador offers an impressive view of the western cliffs of the island, with the young lava platforms below and the recent (and historical) lava flows cascading over the cliffs (Fig. 8.14; best for photography before noon).

Figure 8.14 Recent and historical lava flows of Cumbre Vieja (stop 8.25) overflowing the western cliffs to form extensive coastal platforms. **(a)** General view from the road from Fuencaliente to Los Llanos de Aridane. **(b, opposite)** Oblique aerial view of the 1712 lava flows cascading over the coastal cliff at El Remo. Recent and historical lava flows of Cumbre Vieja (stop 8.25) overflowing the western cliffs to form extensive coastal platforms.

Proceed northwards along the C-832 towards the village of Jedey. The road crosses (km-32 to km-36) the tephritic lavas of Montaña Cabrera–Montaña Faro (15 000 and 18 000 years old respectively), with prominent levees and large **accretionary lava balls**. From km-36 to Jedey the road passes many historical flows corresponding successively to the 1712 and 1585 eruptions, easily recognized by their black colour and distinct levees.

Stop 8.24 The Roques de Jedey and the 1585 eruption   On arriving at the village of Jedey you will see the Roques de Jedey forming a 2 km-long east–west-trending ridge with many phonolite spines and cryptodomes in addition to vents that erupted lava flows at various stages in the eruption of 1585 (Fig. 8.15a).

To have a closer view of this interesting and complex historical eruption, take a left turn and follow the narrow tarmac track that runs up hill into the lava channel of the 1585 flow. The lavas contain many xenoliths of partially fused hydrated phonolite (some of which are vesicular to pumiceous and almost entirely fused) from pre-1585 phonolite bodies, most notably from the Roques de Jedey themselves, which, according to the account of Torriani (1592) were thrust upwards from the ground during the eruption and partially collapsed into the vent that developed at their foot. The juvenile magmas of the eruption were considered to be basaltic but juvenile phonolite bodies that were emplaced as cryptodomes that subsequently broke through the surface were also erupted.

(b)

The pattern of extrusion of younger lavas from fissures in the tops or flanks of older phonolite domes is common in La Palma: in fact, almost every single phonolite dome has acted as a conduit for at least one younger eruption, even though they rise up to 300 m above the surrounding terrain. It seems that the strongly and vertically jointed plugs feeding the phonolite domes are preferred pathways for rising magmas.

You can park here and climb a trail to the top of the volcano and to a

227

Figure 8.15 The Nambroque volcanic complex (most recent eruption 956) as seen from stop 8.28: **(a)** the Roques de Jedey vent of the 1585 eruption (stop 8.24); **(b)** the Nambroque volcanic complex: stop 8.29).

magnificent view of the Roques and the lower slopes of Cumbre Vieja. The ridges on either side of the Roques de Jedey are formed by "whalebacks" or pressure ridges of strongly flow-banded glassy phonolite. These mark where the carapace of older rocks split and fresh magma emerged at the surface. As they did so, the area between was displaced tens of metres westwards and downwards on faults crossing the crest of the ridge just east of the Roques de Jedey. In fact the front face of the dome almost collapsed; if it had done so, a violently explosive eruption could have followed.

Stop 8.25 The coast and coastal platform of Puerto Naos   Past the village of San Nicolás the road passes the 1949 lava flow descending from the Llano del Banco vents, which are just out of sight at the top of the slope. The flow forms a well defined channel on the slope but spread out at its base. The margins of the channel are formed by aa lava levees, but the axial channel (poorly exposed and partly covered by post-1949 alluvium) is mainly formed by pahoehoe lava. A well preserved large lava tube, the Tubo de Todoque, is being prepared for visitors (see Ch. 3 for more on lava tubes).

Drive on down slope towards Puerto Naos and stop at the mirador, part way up the cliff, that provides excellent views of the coastal lava platform to the south as well as of the very young 1949 lava delta (now sadly covered with banana plantations, but with its delta shape still intact despite coastal erosion). On the other side of the road and on the north side of the mirador are exposures of "scree-forming" lavas, here actually composed of a succession of very thin rubbly sheetflows of the 1949 eruption.

Continue north along the coast road, past Puerto Naos. The sequence of cliff forming at this point contains subaerial lavas and scoria for the most part, but exposed at the base of the cliff, with its base close to present sea level (and to sea level *c.* 100 000–120 000 years ago), is the southern rim of a tuff ring composed of yellow palagonitic tuffs. This marks the site of a vent close to sea level at the time of formation, which is also more or less at present sea level; one of the striking features of the Cumbre Vieja is its stability relative to sea level, in marked contrast to the Hawaiian volcanoes discussed in Chapter 1.

The coastal road ends at the fishing village El Remo. This location is on top of the coastal platform of young lavas that has been built up since the end of the last glaciation, at the foot of coastal palaeo-cliffs that have been incised into sequences of older lavas and pyroclastic rocks and are draped by "scree-forming" lava flows. It provides an excellent vantage point to view the principal elements of the stratigraphy of the Cumbre Vieja:

- cliff-forming sequence, cut by the erosion which produced the cliff (120 000–20 000 years ago)
- platform- and scree-forming lavas.

Since the end of the most recent glaciation, lavas descending the cliffs around the coasts of the Cumbre Vieja have merged to form a coastal lava platform that gives the appearance of a raised beach or wave-cut platform but is in fact a constructional feature. Radiometric ages from the platform-forming lavas are all consistent with these lavas having formed in the post-glacial period after sea level had risen to close to its present level. Much of the presently preserved platform on the west coast of La Palma is in fact formed by historical lavas, reflecting the rapid erosion of the platform on this side of the island; even these historical flows have undergone significant erosion in places.

Several scree-forming flows, most associated with the historic 1712 eruption, are visible in the cliffs at El Remo and, although these are very young, they have been removed in places by rockfalls from the cliff, or buried by non-volcanic screes.

The Ruta de Los Volcanes: the summit crest of an active rift
The itinerary begins at stop 8.16. Just as the road begins to climb up the escarpment towards the Cumbre Nueva tunnel, turn right onto a smaller road that leads up to the Refugio del Pilar at the northern end of the Cumbre Vieja ridge. The road climbs up the Cumbre Nueva escarpment, with eastward-dipping lavas and dykes of the Cumbre Nueva rift zone of the Taburiente volcano exposed in the roadcuts.

To the right are fresh aa lavas that were erupted from Montaña Quemada, a spatter cone visible up slope to the southwest and recognizable by a prominent northeast-facing breach in its rim. The eruption that formed Montaña Quemada occurred just before the Spanish occupation of the island (according to Guanche tradition, recorded by the Spanish). A ^{14}C age of 1480 obtained for this eruption is consistent with this.

Stop 8.26 Llanos del Jable–La Barquita The road continues up past Montaña Quemada on Cumbre Nueva rift rocks and then turns east past the back of this spatter cone, passing onto the Llanos del Jable ("plain of sand"; *jable*, corruption of the French *sable*, is applied in the Canaries to sand and lapilli), which is composed of lapilli from the Quemada eruption. These lapilli blanket extensive tracts of the northern end of the Cumbre Vieja.

North of the Llanos del Jable the road continues up over young Cumbre Vieja lavas blanketed with lapilli, most notably those from the Birigoyo volcano, the large recent (6000±3000 years) scoria and spatter cone that forms the northern end of the Cumbre Vieja summit ridge. A mirador on the side of one of the main lava channels from Birigoyo provides excellent views over the El Paso plain and, in the background, the Bejenado volcano and the Caldera de Taburiente (excellent for photography before noon).

The apron of lava flows that extends down the north slope of the Cumbre Vieja partly fill a large basin, bounded to the east by the very steep but deeply eroded lateral-collapse scar of the Cumbre Nueva volcano. Some of these lavas belong to the early cliff-forming series of the Cumbre Vieja volcano, although most of those exposed at the surface are younger than c. 20 000 years old. The town of El Paso is located at the western edge of the basin. Exposures of thick sequences of lacustrine and alluvial sediments in quarries on the floor of the basin indicate that, prior to the growth of the Cumbre Vieja, the basin had previously been partly filled by detritus derived from the Bejenado volcano, the Cumbre Nueva escarpment and, via the deep channel that runs east of Bejenado to La Cumbrecita on the rim of the Caldera de Taburiente, the Taburiente volcano as well. This channel has since been beheaded by deeper incision of the system of barrancos in the Caldera de Taburiente.

Continue up the road towards Refugio El Pilar, crossing and then following the east side of the main Birigoyo lava channel (note large associated lava balls). Shortly after turning east into the pine forest east of this lava flow, turn right onto a forest track (signposted El Gallo) and continue up hill around the side of the El Gallo cone. The track emerges onto a level area with the west side of Birigoyo to the east; here the flank of Birigoyo is formed by spatter intercalated with beds of angular blocks derived from shattered accretionary lava balls, and intact lava balls cover the level ground. Continue up the forest track through pine forests around the flank of La Barquita, another scoria and spatter cone slightly older than Birigoyo. Park on the south side of La Barquita, where the Ruta de Los Volcanes path branches off from the forest road.

Stop 8.27 The 1949 eruption vent and fault system   During the 1949 eruption of the Cumbre Vieja, strong earthquakes were felt west of the ridge, with their epicentre around the village of Jedey. These were associated with the development of gaping fissures and fault scarps along the crest of the ridge for a distance of 3 km. The northern end of the fault system is exposed in the woods southwest of La Barquita. The faults can be reached from the junction of the Ruta de Los Volcanes path with the jeep track by following the contours of the ridge to the south of the barranco that runs west from the junction for about 100 m, then walking slightly down slope in a south-westerly direction for a further 100 m.

About 1 km south, the path passes around the southern edge of Montaña Los Charcos and crosses a barranco by a small bridge. Once off the bridge, a right turn through the wood south of the barranco leads to the Llanos del Agua. This open plain is now covered by hydrovolcanic ash and lithic breccia erupted by explosions from the pit crater of Hoyo Negro in 1949. Gullies at the northern end of the plain expose older yellow ashes (derived from the large explosion crater east of stop 8.29 and also seen in the walls of Hoyo Negro, stop 8.28). These ashes are impermeable: the Llanos del Agua derived its name from the shallow ephemeral pools that formed before 1949 on the surface of this older ash deposit.

On returning to the Ruta de Los Volcanes, the path climbs up onto the ridge east of the Llanos del Agua, from which it is possible to see (cloud permitting) the whole of the northern part of the Cumbre Vieja (Fig. 8.16a). Tenerife is often visible. The northeast-trending scoria cones in the old northeast rift zone can be particularly well seen from this point, as can the precipitous northern face of an old phonolite dome, Nambroque, to the southeast (Fig. 8.15b). The flanks of this dome are draped by later (956, by ^{14}C dating) lavas erupted through fissures in the dome (see also stop 8.24).

Stop 8.28 The 1949 Hoyo Negro hydrovolcanic explosion crater **Warning: the rim of Hoyo Negro is unstable** and the crater is over 100 m deep at this point. Keep at least several metres back from the main rim.

Hoyo Negro is a hydrovolcanic explosion pit that formed in 1949, at the end of the first phase of activity along the summit ridge of the Cumbre Vieja. This had begun with mixed hydrovolcanic and Strombolian activity from the old crater of Duraznero, which is visible to the south (now covered with black to red oxidized spatter from the second phase of activity). This first phase of activity at Duraznero began on 24 June and ended with a violent steam explosion on 6 July. Between the steam explosions at Duraznero and Hoyo Negro, on 8 July a voluminous eruption of lava began from the Llano del Banco fissure (Fig. 8.16b), some 500 m lower down on the western flank of the volcano.

The explosions at Hoyo Negro deposited a thick blanket of lithic blocks (derived from the older rocks in the walls of the growing crater) and fine lithic ash around the crater (Fig. 8.16c). The walls of Hoyo Negro are mainly composed of older scoria cones. The main 1949 faults are visible in a small gully at the northeast corner of Hoyo Negro, cutting through the yellow ash draped by Hoyo Negro ash (see Fig. 8.16c) and as a west-facing scarp on the eastern slope east of the crater, running obliquely southeast across the slope. No other faults are exposed in the older rocks in the lower walls of the crater, which implies that the 1949 fault is the first such fault to have ruptured the surface of the volcano in the period of activity represented by these rocks. Downslope correlations to the west, where rocks from these older scoria cones underlie rocks of the cliff-forming sequence, suggest that this period may be as much as 30 000 years; thus, whatever the significance of the 1949 fault, it is clearly a very unusual feature.

Continue along the Ruta de los Volcanes to the east of Hoyo Negro, or

Figure 8.16 Different views of the Ruta de Los Volcanes itinerary (stops 8.28 and 8.29). **(a)** Aerial view of the Duraznero and Hoyo Negro, vents of the 1949 eruption. **(b)** The Llano del Banco vent, 1949 eruption. (Continued overleaf.)

Figure 8.16 continued Different views of the Ruta de Los Volcanes itinerary (stops 8.28 and 8.29). **(c)** Hydrovolcanic deposits of the Hoyo Negro explosion crater. **(d)** The Duraznero vent and lava lake.

follow the 1949 fault over the crest of the ridge. It terminates about 80 m to the southeast of the point where it crosses the path again, but at this southern end it forms an impressive open fissure, the western side of which is displaced downwards relative to the eastern side by 2–3 m. The wooded upper surface of the Nambroque phonolite dome to the east is punctuated by high spines of phonolite, extruded from the surface of the dome.

Stop 8.29 The 1949 Duraznero eruptive vent   The trail ascends to the rim of the Duraznero vent, the site of the early (24 June to 6 July) activity during the 1949 eruption. Cutting the north rim of the Duraznero crater and extending across its floor is a line of spatter vents along an eruptive fissure, with tree moulds clearly visible (Fig. 8.16d).

East of the fissure the lavas fill an old depression (an old and mostly filled explosion pit, similar to but larger than Hoyo Negro) forming a spectacular lava lake from which lava flows down slope towards the eastern coast (the 1949 flow mentioned in stop 8.22). The surface of the lava lake is locally deformed by pressure ridges and occasional tumuli. Gas blisters and drain-back crevasses are also seen.

Stop 8.30 Volcán de San Antonio and the 1677 Fuencaliente eruption   The trail climbs the Deseada (1949 m), the highest volcano of the Cumbre Vieja, and starts the continuous descent towards Fuencaliente, 8 km down hill. Past the Cabrito phonolitic dome and the tephritic cones of Montaña Cabrera and Montaña Faro, the trail passes to the right of the pre-Martín group of cinder cones, topped with hydrovolcanic deposits and the vents of the 1646 eruption forming a lineation of conelets feeding a wide lava channel (Fig. 8.17a).

Close to Fuencaliente and past the football field built inside the crater of a cinder cone, a turn of the trail opens the view of the southern end of the Cumbre Vieja Rift. The San Antonio scoria and spatter cone is one of the largest in the entire Cumbre Vieja (Fig. 8.17b), rising well over 200 m above its surroundings. It is also notable for the thick sequence of indurated hydrovolcanic ashes and ash-rich breccias, with low-angle cross bedding indicative of surge development, which occurs on its western rim and points to violent explosions at the end of the eruption in which it formed. Nevertheless, local tradition (and tourist postcards) identify San Antonio as having formed in the eruption of 1677 (which, incidentally, did *not* start on St Anthony's Day). Had it done so, the explosions at the end of the eruption would have had a devastating effect upon Fuencaliente, whereas contemporary accounts indicate that the eruption was comparatively mild in its effects.

Several lines of evidence indicate that the San Antonio cone is in fact

Figure 8.17 The San Antonio cinder cone and the eruptions of 1646 and 1971 (stops 8.30 and 8.31). **(a)** Aerial view of the San Antonio cone and the upper vent of the 1677 eruption of Fuencaliente. **(b)** Aerial view of the upper vent (Martín) of the 1646 eruption of La Palma. **(c, opposite)** Aerial view of the vents and flows of the Teneguía volcano eruption in 1971, the last eruption in La Palma and the Canary Islands.

much older than 1677. Lavas erupted from vents in Fuencaliente itself and which pre-date the Montaña del Fuego lavas (3200 years old) pass around the cone and are affected by its presence (see Fig. 8.17b); and no occurrences are found of the hydrovolcanic ash unit on top of these flows. In addition, pre-Hispanic (i.e. pre-1500) archaeological remains of the Guanche culture occur on the western side of San Antonio, and a cone named Montaña San Antonio and in the correct location appears on a map draw by Torriani dating from the end of the sixteenth century, 85 years before the 1677 eruption. The evidence of the San Antonio being an old volcanic cone not related to the 1677 eruption (the Fuencaliente eruption) is now on display in the new visitor centre.

The 1677 eruption of Fuencaliente formed a Strombolian cone partly attached to the northern flank of the San Antonio cone (the crater partly occupied by the car-park of the visitor centre) and a group of vents on the southwest side of the cone, the source of many lava flows, as can be seen from the rim of the San Antonio cone spreading west and south to the coast.

The scoria and lava produced in the 1677 eruption are also notable for their high content of metasomatized peridotite and MORB gabbro xenoliths from the old oceanic lithosphere beneath La Palma, as well as younger pyroxenite and alkali gabbro xenoliths. In contrast, the San Antonio cone rocks contain only pyroxenite xenoliths, some of which (in the late hydro-volcanic deposits) are of impressive size.

Stop 8.31 Lower vents of the 1677 Fuencaliente eruption and Teneguía volcano
Drive along the track leading to the car-park back to tarmac road. Turn left
at the first junction, towards Las Indias, but turn sharply left again at the
entrance to Las Indias onto a rough track that leads around the western
base of the San Antonio cone. Just past the Roque de Teneguía phonolite
spine, a parking bay overlooks the lower southwest vents of the 1677 eruption. These vents, which are aligned along a northwest–southeast-trending fissure, fed several lava flows that form an extensive coastal platform
(and buried the Fuente Santa, a famous medicinal hot spring). There is a
strong contrast between these small hornito-like lower vents and the
Strombolian scoria cone that formed at the upper vent. This same vertical
or topographical variation is seen in many other multiple-vent eruptions
in La Palma, as in the 1585 Jedey eruption and the Llanos del Banco vents
of the 1949 event. It appears to reflect upward migration of gas bubbles
within the feeder dykes, and the consequent eruption of gas-enriched
magma at the uppermost vents of each eruption fissure system.

Continue south and east along the gravel track to the south of the San
Antonio cone, to a viewpoint above the Teneguía scoria and spatter cones,
which formed in the 1971 eruption, the most recent of Cumbre Vieja
(Fig. 8.17c). A group of vents along fissures trending north–south, formed
scoria and spatter cones and fed mainly basic lava flows. The lavas contain
fused phonolite xenoliths, possibly another case of an eruption controlled
by a phonolite dome that acted as a pathway to the surface (see stop 8.24).
These flows were mainly directed to the west and south (see Fig. 8.17c) and
created a significant coastal lava platform, partly overlying the 1677 lava
platform.

Fumaroles inside and along the crest of the main cone of the Teneguía
volcano still give off residual vapour evident in a locally strong sulphur-
ous odour. Temperatures over 210°C were measured in the crest of the
main cone in 1985 and have declined since to less than 100°C.

Chapter 9

El Hierro

The island of El Hierro is the emergent summit, with an area of 280 km², of a volcanic shield that rises from the sea floor, 156 million years old, from a depth of 3700–4000 m below sea level. The most notable feature of the island is its tri-lobe shape, with three convergent ridges of volcanic cones (a triple-arm rift system) separated by wide embayments (Fig. 9.1).

El Hierro is one of the best places to observe the close relationship between rift zones and giant gravitational or lateral collapses. Giant gravitational flank failures were first proposed by Moore (1964) to explain the great seacliffs of the Hawaiian Islands. However, confirmation of the

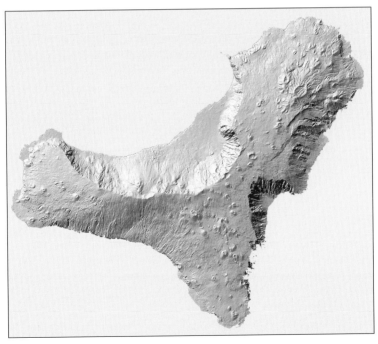

Figure 9.1 Shaded relief image of El Hierro, with the main geomorphological and tectonic features (image GRAFCAN).

hypothesis waited until the 1980s, when the debris of the landslides (including blocks 10 km or more across) was mapped off shore, using sonar instruments. This is because Hawaiian lateral collapses remove huge portions of the submarine parts of the volcanic edifices (up to several thousand km^3), but mostly have headwalls right on the coast and so are difficult to observe. Conversely, giant landslides in El Hierro (and in La Palma) are comparatively small (typically 100–200 km^3) and develop shallow fault planes at depths of 1–2 km that mostly remove the emergent part of the island volcano. The collapse embayments, nested in the confining rifts, form a clearly discernible and spectacular landscape. The corresponding submarine debris-avalanche deposits have been clearly identified from swath bathymetry, seismic reflection and side-scan sonar data.

As already mentioned in Chapter 8, La Palma and El Hierro form a dual line of island volcanoes at the western end of the hotspot-related Canarian chain. El Hierro is the youngest of the Canaries: the oldest surface lava dated is 1.12 million years old. However, unlike La Palma, which has had seven eruptions in the 500 years of recorded history, the records in El Hierro show only a seismic crisis in 1793. This was interpreted as related to a volcanic eruption (Lomo Negro volcano), but most probably corresponds to a shallow intrusion of magma or a nearby submarine eruption, since there is no mention whatsoever of observed volcanic activity from El Hierro or La Palma. The youngest eruption is, probably, a vent in the Montaña Chamuscada–Montaña Entremontañas volcanic group near the village of San Andrés on the central plateau of El Hierro, dated in 500 BC.

Why has La Palma been much more active than the younger El Hierro during the Holocene? Both islands may have some connection in their magma source, at least in the past 700 000 years (when the southward migration of volcanism started in La Palma). Periods of intense volcanism on one island coincide with periods of relative inactivity on the other. A period of intense activity on one island culminates in a giant lateral collapse, which is followed by a switch in the location of the most intense volcanism to the other island. One can speculate that the switch may be caused by the unloading effect of the collapse, which would place the rebounding lithosphere beneath into horizontal compression. Does that imply that a future intensification of volcanism in El Hierro will occur only after the collapse of the Cumbre Vieja volcano of La Palma?

El Hierro developed during the Quaternary by the superposition of three main volcanic edifices separated by major tectonic events (the gravitational collapses):

1. Tiñor volcano (the seamount stage does not outcrop in El Hierro)
2. El Golfo volcano
3. Rift volcanism.

The El Hierro volcanoes, typical of oceanic islands, present a simple geochemical evolution, possibly in consonance with the island's rapid growth. The first subaerial eruptive cycle, which produced the Tiñor edifice, is characterized geochemically by relatively primitive basalts to trachybasalts and tephrites. The lavas of the ensuing El Golfo volcano are geochemically more evolved, with highly differentiated trachytes at the top of the sequence. However, compositional overlap is evident in the basic and intermediate-composition components of the Tiñor and El Golfo edifices. The last eruptive sequence, consisting of the products of widespread eruptions located along the triple-arm rift system, is characterized by a range of alkaline picrobasalts, basanites and tephrites. The Rift lavas appear to be slightly more alkaline and silica undersaturated than those of the first two edifices. Trace-element data indicate that the rocks of each edifice can be related by fractional crystallization processes leading to a systematic enrichment in incompatible elements.

The simplified geological map of Figure 9.2 is based on geological and geochronological studies that defined those volcanic edifices and their main volcanostratigraphic units.

The Tiñor volcano

The Tiñor volcano formed the first stage of subaerial growth of El Hierro. Its present outcrop is confined to the northeastern flank of the island and inside the Las Playas embayment (Fig. 9.2). The Tiñor volcano developed very rapidly and there is no consistent compositional variation with time that can be mapped in the field. However, there are some differences between units that may reflect the morphological evolution of the developing edifice:

- a basal unit of relatively thin steep-dipping flows, probably corresponding to the initial stages of growth of the volcano, with steep flanks
- an intermediate unit of thicker lavas, that progressively trend to subhorizontal flows in the centre of the edifice, probably reflecting the lower slopes of this mature stage of growth of the volcano
- a group of emission vents with well preserved craters (the Ventejis volcano group), and lavas filling valleys and canyons carved into the older rocks.

The flows of the Ventejis Group are very easily identifiable because they are rich in xenoliths, probably because they correspond to terminal, explosive stages immediately preceding the collapse of the northwestern flank of the Tiñor volcano.

Cross sections 2–4 in Figure 9.3 show the relative stratigraphic position

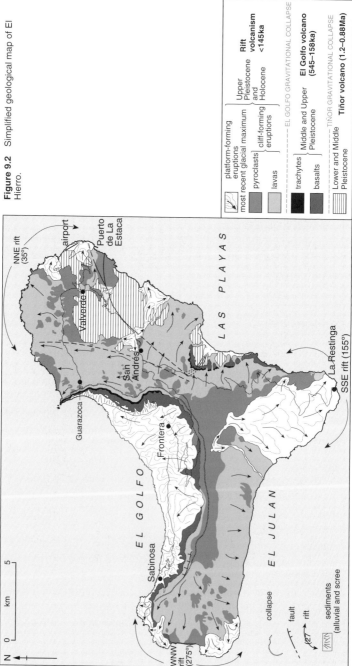

Figure 9.2 Simplified geological map of El Hierro.

Legend:

platform-forming eruptions — most recent glacial maximum

cliff-forming eruptions — pyroclasts / lavas — Upper Pleistocene and Holocene

Rift volcanism <145ka

———— EL GOLFO GRAVITATIONAL COLLAPSE

trachytes / basalts — Middle and Upper Pleistocene — **El Golfo volcano (545–158ka)**

———— TIÑOR GRAVITATIONAL COLLAPSE

Lower and Middle Pleistocene — **Tiñor volcano (1.2–0.88Ma)**

collapse
fault
rift
sediments (alluvial and scree)

N

0   km   5

NNE rift (35°)
airport
Puerto de La Estaca
Valverde
San Andrés
Guarazoca
La Restinga
SSE rift (155°)
Frontera
Sabinosa
WNW rift (275°)

LAS PLAYAS
EL GOLFO
EL JULAN

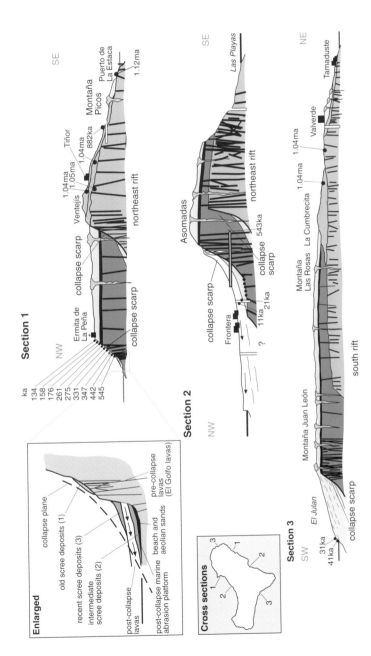

Figure 9.3 Different cross sections of El Hierro (see lower inset map).

of the Tiñor and the subsequent volcanoes. The limited extent of the Tiñor volcano towards the south and west is evident in the sections 1 and 3 of this figure.

The El Golfo volcanic edifice

After the Tiñor volcano collapsed, a new volcanic edifice (El Golfo volcano) developed, filling the northwest-facing collapse embayment and finally spilling lavas towards the east coast overlying the Tiñor volcano (see section 3 in Fig. 9.3).

The El Golfo volcano developed entirely in the Brunhes period. One of the lowest lavas filling the El Golfo embayment gave an age of 545 000 years. Therefore, an important break in the activity of El Hierro may have taken place between the Tiñor and El Golfo volcanoes, probably coinciding with the period of rapid growth of the Cumbre Nueva volcanic rift zone in the nearby island of La Palma.

The radial dips of the lava flows indicate that the El Golfo edifice was centred near the town of Frontera, inside the later collapse embayment. The summit region may have been as much as 2000 m above sea level. Two sub-units may be identified in this volcano from morphological differences and from the local development of unconformities:
- a basal unit, predominantly composed of Strombolian and Surtseyan pyroclastics (cinder cones and tuff rings), with subordinate lava flows
- an upper unit predominantly composed of lava flows.

The lower unit is cut by many dykes that form swarms trending northeast, east-southeast and west-northwest, which match the present volcanic vent systems and indicate that a triple-rift system was an important feature of the El Golfo edifice. In contrast, the relatively few exposed feeder dykes in the Upper El Golfo unit indicates that the lavas that make this up mainly originated from near the summit of the El Golfo volcanic edifice, although some flank vents are spectacularly exposed in the El Golfo cliffs. This central concentration of volcanic vents is in marked contrast to the younger Rift-series volcanism. The upper El Golfo unit is topped by several differentiated (trachybasalts, trachytes) lava flows and block-and-ash deposits. We interpret these volcanic differentiates as the terminal stages of activity of the El Golfo volcano, prior to the establishment of the Rift volcanism.

The duration of the growth of El Golfo volcano can be estimated at about 360 000–380 000 years, as indicated by the lower age of 545 000 years and the age of the trachytic lavas (176 000 years) in the collapse scarp section (see section 1 in Fig. 9.3).

The Rift volcanism

Although a triple rift system was present in the El Golfo volcano and may also have controlled the development of the northeast-trending Tiñor volcano, we define the Rift volcanism as the late stage of growth of the island when the three arms of the Rift have been simultaneously active without a central vent complex, as appears to have been present during the growth of the El Golfo activity. The Rift-series lavas are broadly concordant on El Golfo lavas in much of the island, but striking local unconformities are present, especially near the old coastlines. As a result of this wide distribution of vents, a relatively thin mantle of basic lavas has covered much of the island. These lavas have mainly filled the El Julan collapse embayment and partially filled the El Golfo embayment (Fig. 9.2; see sections 2 and 3 in Fig. 9.3; also discussion of island structure below).

The maximum age these eruptions is constrained by the differentiated emissions topping El Golfo volcano (158 000 years). Radiometric (K/Ar and ^{14}C) ages from 134 000 years to, at least, 500 BC, indicate that activity in this late volcanic stage is continuing, although the eruptive rates are relatively moderate. As yet the Rift-series volcanism has not, therefore, produced a well defined volcanic edifice comparable to the Tiñor and El Golfo volcanoes. This may be the consequence of the migration of volcanic activity to the nearby island of La Palma, where the very active Cumbre Vieja volcano has developed in this period.

Giant lateral collapses of El Hierro

The Tiñor giant collapse

Probably soon after the late explosive episodes of the Tiñor volcano activity (the xenolith-rich Ventejis eruptions about 882 000 years ago), the volcanic edifice collapsed towards the northwest, producing the first giant landslide of El Hierro. This collapse may have removed more than half of the volume of the subaerial part of the Tiñor edifice (Figs 9.1, 9.2).

The evidence for this giant collapse is shown in cross section 4 of Figure 9.3, which shows a galería excavated in the El Golfo embayment scarp that extends towards the lavas of the Tiñor volcano. At the end of the tunnel, the El Golfo lavas of 543 000 years and normal (Brunhes) polarity are at the same level as gently eastward-dipping lavas more than 1.04 million years old and of reverse polarity (Matuyama pre-Jaramillo), which are exposed at the surface and in galerías to the east.

The El Julan giant collapse

The El Julan collapse was identified in 1991 and considered older than the El Golfo collapse. The lack of outcrops of the collapse scarp make the dating of this event from onshore evidence difficult. Water galerías in the El Julan collapse embayment cross part of the filling lavas, belonging to the Rift volcanism. This constrains the minimum age of the collapse to about 150000 years. The El Julan collapse, which destroyed the southwestern flank of the El Golfo volcano, probably occurred when this volcano was well developed.

The San Andrés aborted giant collapse

The San Andrés collapse is both an excellent example of these rare tectonic structures and a rich source of information on the development of catastrophic collapses in general. It appears as a fault with some unique features evident at outcrop that reveal its origin (see stop 9.4). It is relatively young but inactive and it developed between 545000 and about 261000–176000 years ago. It formed along the flank of the steep-sided northeast rift of El Hierro and is bounded by a discrete strike-slip fault zone at the up-rift end, closest to the centre of the island.

This geometry differs markedly from that of collapse structures on stratovolcanoes, but bears some similarities to that of active fault systems on Hawaii. Although the fault has undergone little erosion, **cataclasites**, which formed close to the palaeo-surface, are well exposed. These cataclasites are among the first fault rocks to be described from volcano lateral-collapse structures and include the only **pseudotachylytes** to have been identified in such structures to date. The structure of the fault-rock outcrops and their implications for collapse mechanisms are discussed in the location descriptions (stops 9.4 and 9.5).

The well developed topographic fault scarp associated with the San Andrés fault system led to the hypothesis that it was an active incipient collapse structure and, therefore, a major natural hazard. However, the age relationships of the faults to dated lavas and other volcanic rocks point to a different conclusion (stops 9.3 and 9.4) and show that it is not a hazard. It is an old and inactive structure that is unlikely to be reactivated. After the aborted collapse, the El Golfo giant landslide occurred without reaction of the San Andrés fault.

The El Golfo collapse (or collapses?)

The El Golfo embayment is perhaps the most spectacular feature of El Hierro. It is some 15 km across from Roques de Salmor to Arenas Blancas, extending about 10 km inland from these points, and its headwall is still more than 1.4 km high in places. Taking into consideration the likely original height of the El Golfo edifice (about 2000 m) the probable volume of subaerial material removed in the formation of the embayment is at least 120 km^3. In addition, the available bathymetry indicates that a similar volume has been removed below sea level.

Formation of the El Golfo embayment by catastrophic lateral collapse was first proposed on the basis of the discovery of a giant debris-avalanche deposit off shore to the north. In favourable conditions, the sliding into the ocean of such a huge volume of rocks may have produced a huge **tsunami**, probably affecting the remainder of the Canaries and beyond.

The age of the El Golfo embayment is still problematic. Some authors proposed a single collapse that occurred 13000–17000 years ago, based on the correlation of the collapse debris-avalanche deposits found off shore to the north with a turbidite in the Madeira abyssal plain.

This offshore information strongly conflicts with onshore evidence for the age of the embayment (stops 9.6–9.18, especially 9.6, 9.14 and 9.15). This evidence suggests that the subaerial embayment may have formed soon after the emplacement of the lavas at the top of the cliff sequence (134000 years old) during the previous interglacial period. A possible means of reconciling the contradictory onshore and offshore evidence is to postulate the occurrence of two lateral collapses, one mainly subaerial and occurring at about 100000–130000 years ago, and the other affecting the seaward parts of the lava platform built up within the embayment, and also the submarine slope of the island down to considerable depths. This last event, which may have taken place between 17000 and 9000 years ago, would have produced the megaturbidite *b*. This sequence of events is summarized in Figure 9.9b.

In summary, the volcanic history of El Hierro clearly exemplifies the intense interaction between volcano growth and lateral collapse episodes in the very early stages of subaerial development of an oceanic island. The present-day island of El Hierro is therefore a fraction of the three successive volcanoes accreted onto earlier, partially destroyed edifices. The present subaerial volume of the island is probably less than a third of the volcanic products erupted, which gives an idea of the difficulties encountered in these islands in evaluating magma production and eruptive rates.

Logistics on El Hierro

Most of the recommendations described for La Palma in Chapter 8 apply equally to the island of El Hierro, only with more stringent limitations. El Hierro is the least populated and least visited island of the Canaries. As a result, it has the poorest transport connections of all the Canaries, although several scheduled flights are available daily from Tenerife and Gran Canaria, as well as a cheaper ferry from Santa Cruz de Tenerife and Los Cristianos (south Tenerife). However, the airport is occasionally closed for days when westerly winds ("viento de montaña") blow down the mountain and laterally into the airport. Only small aircraft operate to El Hierro. The ferry service in the small and poorly protected Puerto de La Estaca can also be interrupted by the weather, so it is best not to make the island the last stop on a multi-island tour before returning home.

There are few hotels in the island, so book well in advance. However, there are small pensiones and apartments in Valverde, Frontera–Tigaday and La Restinga. The casas rurales, available throughout the entire island, are a very good choice, many of them close to the areas of greatest geological interest (www.el-hierro.org). The island is small and distances not a problem. Public transport is not good and taxis are expensive. Hire cars are the best choice, since they are cheap and plentiful, but book well in advance (particularly if you plan to rent a vehicle with four-wheel drive).

The weather in El Hierro is seldom a problem, except for photographic purposes. The windward part of the island,[*] the most Atlantic of the Canaries is very frequently under constant trade winds, rendering the climate very mild throughout the year but favouring the presence of the mar de nubes ("sea of clouds") attached to the cliffs of El Golfo embayment and the north and northeast flanks of the island. Conversely, the southwest coast (El Julan) is almost constantly clear and the sea calm (mar de las calmas: "sea of calm"). Occasionally, all the Canaries are influenced by winds from the African desert, the tiempo sur ("south weather"): extremely hot, dry and dusty, with very low visibility.

As advised in Chapter 8, keep well back from the very unstable edges of the coastal cliffs. Take precautions against excess exposure to the Sun and against dehydration. A cellular phone, with satellite coverage for the emergency number 112, can be a safety bonus.

About 60 per cent of the island is protected by law. The park of Frontera alone protects nearly the entire western half of the island. Other parks and reserves include the cliffs of the collapse embayments of El Golfo and Las Playas, and most of El Tiñor volcano. There is no national park in El Hierro. However, if you plan to sample, ask for information and apply for a permit in advance from the Department of the Environment of the Canarian government (www.gobcan.es/medioambiente) to avoid embarrassing situations in the field or when checking out at the airport.

Very precise topographic maps (1:25000 and 1:5000) and aerial photos,

---

[*] The worldwide reference standard or prime meridian was located at the Faro de La Orchilla, on the west coast of the El Hierro, before being established at the Royal Astronomical Observatory at Greenwich.

showing every road and track, are available from the Instituto Geográfico Nacional (www.cnig.ign.es) and the Canarian government (www.graf-can.rcanaria.es). Geological 1:25 000 maps of El Hierro by the Instituto Geologico Nacional (igme@igme.es) are in press.

Four itineraries are proposed (Fig. 9.4) for the observation of:

- the main features of the Tiñor volcano and fossilized collapse escarpment (stops 9.1–9.3), and the San Andrés fault system (stops 9.4–9.5)
- the rim of the El Golfo collapse scarp and the general features of the collapse embayment from identified vantage points (stops 9.6–9.11); clear conditions are needed to get the best of these views
- the main volcanostratigraphic units of the El Golfo volcano, and the evidence to date approximately the occurrence of this gravitational landslide (stops 9.12–9.15)
- the recent volcanism of the rifts (stops 9.16–9.21).

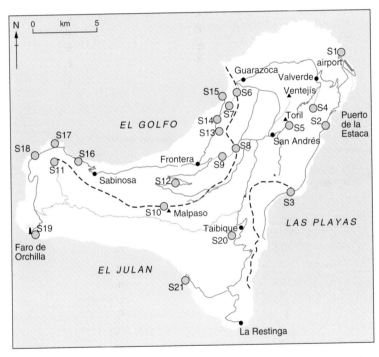

Figure 9.4   Sketch map showing the main roads and the stops of the geological itineraries.

The Tiñor volcano

As mentioned earlier, the Tiñor volcano outcrops only in the northern part of the island, in the area of the airport, the harbour of Puerto de La Estaca and the island's capital town. This is therefore the best itinerary to follow on the day of your arrival, and also because it covers the oldest geological formations.

The southeastern flank of the Tiñor volcano

About 1 km after leaving the airport, turn right at the first junction to El Tamaduste, about 1 km distant.

Stop 9.1 El Tamaduste The village and beach of Tamaduste are sheltered in a small bay formed at the foot of a palaeo-cliff cut in the Tiñor volcano lavas. Several cinder cones appear interbedded and the entire formation is covered with a layer of basaltic lapilli from the recent vent of Montaña La Cancela, at the top of the cliff. Lavas from this vent formed a coastal platform, creating a narrow sea entrance with a small beach.

Just outside the embayment the road cuts an apparently recent cinder cone overlain with xenolith-rich lava flows (from the Ventejis volcano up hill). This is a very good example of how deceptive appearances can be in determining the ages of volcanic eruptions. A recent cinder cone or lava flow deposited in a very humid environment (in the seaward flanks or under the water table) may appear older than a much older vent emplaced in drier areas. The apparently recent cone (black, fresh cinder with clean vesicles) underlies lavas of reverse (Upper Matuyama) magnetic polarity corresponding to the Ventejis volcano, a late eruptive vent of the Tiñor volcano, dated at 0.88 million years old. A closer inspection reveals truncated dykes (also of reverse polarity) cutting the cinder cone.

On returning to the junction, take the road to the left, to Puerto de La Estaca and Las Playas.

Stop 9.2 The southeastern flank of the Tiñor volcano The road (TF–911) continues along the flanks of cinder cones (Picos, Ribera) and, past km-2, on lavas of the Tiñor volcano to Puerto de la Estaca and to Bahía de Tijimiraque. At the entrance to this bay there is a small outcrop of pillow lavas at sea level, which is evidence of the absence of subsidence in the island (as in the entire Canarian archipelago, as discussed in Ch. 1).

Past the bay, the road crosses a coastal platform of lavas from the Montaña Chamuscada vent (RL on Fig. 9.5a) flowing inside barrancos carved in steeply dipping lavas (SDL) of the Tiñor volcano. At the top of the sequence outcrop horizontal lavas (HL) of the San Andrés plateau.

Figure 9.5 **(a)** Rift lavas of Montaña Chamuscada (near San Andrés) cascading over the northeast flank of El Hierro (stop 9.2). The recent lavas flow in barrancos cut in the steep lavas of the Tiñor volcano (SDL) and the horizontal lavas (HL) filling the Tiñor embayment to form a perched central plateau. **(b)** Panoramic view of Las Playas (stop 9.3), with the Roque de Bonanza erosive remnant in the foreground. The cliff is formed by lavas of the Rift series (RL) and the trachytic and basaltic lavas of the El Golfo volcano (T and Garafía volcano), topping the Tiñor volcano lavas (Tiñor volcano). The Rift lavas and scree (SD) show different depositional angles.

Continue through the road tunnel; where it ends, the view opens onto the Las Playas embayment.

Stop 9.3 Las Playas   The Barranco de Las Playas is a large, roughly triangular topographic embayment, open to the sea to the southeast and bounded by cliffs up to 900 m high on the other sides.

In the foreground is the Roque de La Bonanza, a remnant in the form of an arch of an eroded vent. In the escarpment at the background are outcrops, from top to bottom, the Rift lavas, the light-coloured trachytes and basalts of the El Golfo volcano, and, at the footwall, the Tiñor volcano lavas.

A series of what from afar look like scree deposits drape the footwall of the escarpment. However, upon closer inspection the steep Rift lava flows cascading down the cliff can be easily separated from the lower-angle scree deposits.

This embayment used to be interpreted as a small-scale collapse structure. However, its proportions in plan are different from those of other known collapse embayments in the Canaries, which are typically longer (along the coast) than they are wide (perpendicular to the coast and to the rift zones at their heads), or else of sub-equal proportions, and have a discrete headwall; they are thus quadrilateral rather than triangular in plan.

Detailed mapping indicates, furthermore, that the Las Playas embayment lies along the line of a strike-slip fault system, trending northwest–southeast, which bounds the southern end of the San Andrés fault system (Figs 9.1, 9.2). No faults, apart from superficial fissure systems parallel to and close to the tops of the present-day seacliffs, occur to the south of Las Playas. In contrast, a swarm of northwest-trending vertical faults cut the older rocks at the apex of the embayment, and southeast-facing normal faults are visible in the northern wall. We therefore infer an asymmetric geometry for the San Andrés fault system, with the main rift-parallel normal faults bounded by an arcuate set of splay faults with oblique slip at the down-rift end (Figs 9.1, 9.2) and by a discrete set of strike-slip faults at the up-rift end, closest to the intersection of the three volcanic rifts of El Hierro. This has some similarities in geometry (although not in kinematics) to the south rift of Kilauea (Hawaii): the main normal fault is analogous to the Hilina faults, and the strike-slip faults to those near Kilauea, which link the south and east rifts of that volcano.

We therefore infer that, rather than being a discrete collapse embayment, the Las Playas is primarily a giant barranco system (similar to, although much smaller than, the Caldera de Taburiente) that has been eroded along a weak zone formed by the strike-slip fault system bounding the southern end of the San Andrés structure. This fault zone appears to have performed a function similar to that of the northern boundary scarp of the Cumbre Nueva collapse structure in localizing erosion in this area, although it is possible that a much smaller initial collapse structure, perhaps produced as a superficial feature on top of the sliding San Andrés fault block, may also have contributed to the localization of erosion in this area. We emphasize again how rapidly deep erosion can occur in the Canaries once a drainage system has been fixed in position by volcano structures.

Exposures in the cliffs of Las Playas also provide critical age constraints on the age of the San Andrés fault system. The faults at the head of the Barranco de Las Playas cut the two lowest units in the cliffs, the mainly scoriaceous Tiñor series and the lower lavas of the El Golfo formation (545 000–442 000 years ago), but are truncated at the remarkably planar unconformity at the base of the lavas of the upper sequence (not dated here, but 261 000–176 000 years old elsewhere in the island). This unit can be traced around the entire rim of the Barranco de las Playas, as can the overlying Rift-series lavas, without offsets; both units can be seen in the eastern cliffs. In addition, scree-forming Rift lavas, two of which have been dated at 44 000±3000 and 145 000±4000 years ago, occur intercalated with the screes in the lower parts of the barranco. In the context of problems associated with the age of the El Golfo embayment (stops 9.4 and 9.5), it is noteworthy that the 145 000-year-old lava is about 70 000 years older than a lava that is apparently part of the "cliff-forming" sequence on the north rim of Las Playas. These observations indicate that, on the basis of its crosscutting relationships to dated rock units, the San Andrés fault system formed between about 545 000 and 261 000–176 000 years ago.

The San Andrés fault system
The San Andrés fault system runs along the eastern side of El Hierro (Figs 9.1, 9.2). During the early 1990s it was the cause of great concern as a potentially active incipient landslide structure (on the basis of the well developed fault scarp and well exposed fault rocks) that might fail and trigger a giant lateral collapse and associated tsunami, with catastrophic consequences. Field and geochronological evidence indicates that it is in fact an aborted lateral collapse structure that has been inactive for as much as 250 000 years. However, it is a unique structure, at least in the Canarian archipelago and possibly in the world, and is a potentially unique source of information on the structure and mechanics of the initial stages of failure, leading under other circumstances to giant lateral collapses.

Stop 9.4 Outcrops of the San Andrés fault at the barranco de Tiñor   From the airport junction, proceed to Valverde along road TF-911. Where the road leaves Valverde towards San Andrés and Frontera (TF-912), turn left onto the track leading down to Barranco de Tiñor.

The col in which this junction lies is on the line of the main northeast–southwest-trending San Andrés fault, just where it begins to curve around to a more easterly trend (Fig. 9.2). At this point the scarp has been almost completely eroded; to the north the fault is concealed under younger lavas. Here the fault juxtaposes the upper parts of Brunhes-epoch scoria cones (Montañas Picos and Riviera) in the hanging wall to the southeast against lowermost Brunhes-epoch scoria and Matuyama-epoch lavas in the footwall to the northwest.

The amount of offset at this point is unclear because of the lack of good stratigraphic markers and the intense erosion that affected the area both before and after fault movement. However, an impression can be gained from the height of the Ladera de Gamonal, a steep slope formed by the fault escarpment, to the south and west: this is 200–300 m high along most of its length. The sequence of lavas and scoriaceous pyroclastic rocks visible in this escarpment belongs entirely to the Tiñor volcanic edifice.

Continue on foot along the road (which rapidly degrades into a jeep trail) to the southwest for about 15 minutes until the north rim of the Barranco de Tiñor is reached.

The Barranco de Tiñor is the deepest of several barrancos cut into the San Andrés fault scarp. Like several of the others it contains young lava flows not offset by the fault scarp, visible here as a linear crag formed by indurated fault rocks along the foot of the main cliff of the fault scarp (Fig. 9.6). Estimation of the offset on the fault is complicated, but may be of the order of 300 m across the steeply dipping fault surface.

The fault itself is well exposed along the base of the scarp in the form of a discontinuous but strikingly linear ridge up to several metres high and with a steep southeast-facing surface composed of the indurated rocks in the fault zone. This indurated zone is less than 0.5 m thick. The best exposures of all are along a recently excavated roadcut on the north side of the barranco. The wall of this cut is formed by the spectacularly grooved fault plane itself; the footwall and hanging-wall rocks adjacent to the fault plane are exposed in an adjacent section of the roadcut and in the barranco.

At this locality, both footwall and hanging-wall rocks are mainly basaltic lavas of the Tiñor series, with a few thin intercalated lapilli beds and soil horizons. Lava flows, red soil and lapilli beds, and even individual flow lobes, can be traced to within a metre of the fault surface on both sides. The footwall rocks appear undeformed outside the fault zone. Examination of the extensive outcrops in the barrancos on either side of the fault has confirmed that there is little or no faulting outside the fault zone. The total thickness of the fault zone appears to be less than 1 m, most of which is

Figure 9.6 View of the main outcrop of the San Andrés fault in the Barranco de Tiñor (stop 9.4).

formed by a coarse clast-supported incohesive breccia, best seen in the footwall, where the barranco bed cuts the sheet of indurated fault rock, and in the hanging wall at the northern end of the exposure, although isolated fragments still adhere to the fault surface.

Embedded within this breccia is a continuous sheet, 30 cm or so thick, of finer-grain cohesive fault rock. On the eastern and upper surface of this indurated breccia is a thin (0.5–1 cm) layer of strongly indurated porcellaneous dark purplish-grey rock that forms the fault surface itself. This rock is a matrix-supported microbreccia, with relatively few small identifiable lithic clasts (only a few millimetres across) set in the porcellaneous matrix. Again, there is no foliation or shape fabric in this rock. This part of the fault plane in particular shows well developed and delicately preserved structures ranging from millimetre-scale grooves, through toolmarks left by rigid blocks within the fault breccia, to metre-scale undulations. These structures have been used to determine the slip direction on the fault surface at this and other locations: although there are local variations, the overall slip direction is southeastwards, towards the sea, throughout the San Andrés fault system and there appears to have been very little internal deformation within the sliding block.

Thin-section petrographic examination of these rocks indicates that the finer-grain rock appears to be a frictional melt rock or pseudotachylyte, produced by frictional heating because of large and rapid movements on the fault surface.

The implication that the bulk of the movement on the San Andrés fault system represents a single slip event is consistent with thermal calculations of the amount of frictional heating required to produce the melting on the fault surface. At the near-surface depths of these exposures, slips of the order of hundreds of metres are required to produce melting; the only other settings in which frictional melt rocks have been found in near-surface faults are the basal detachments of large non-volcanic landslides such as Kofels in the Alps and Langtang in the Himalayas. Therefore, it appears that the bulk of movement on the San Andrés fault took place in a single slip event, with movement of some hundreds of metres in, at most, a few minutes.

The critical evidence for present-day inactivity of the San Andrés fault is also exposed in the bed of the Barranco de Tiñor. A recent lava flow descending the barranco is exposed a few metres west of the intersection of the fault rock with the bed of the barranco, and again a few metres to the east, although it has been removed by erosion at the intersection itself.

There is no evidence for offset of this lava flow across the fault. Similar relationships are evident in the smaller barrancos to the north (Fig. 9.2) and at stop 9.5. Although these particular flows have not been dated, on the basis of the mapping they belong to the Rift series lavas and could be as much as 150 000 years old.

To the south, the bend in the track as it emerges from the Barranco de Tiñor offers a convenient point to view the main San Andrés fault scarp, the Ladera de Gamonal (Fig. 9.7).

Throughout this vista, the San Andrés fault system is primarily composed of a single fault with a single dominant slip surface. This is an

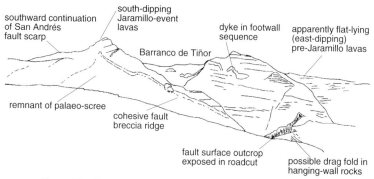

Figure 9.7 Sketch of the main outcrops of the San Andrés fault system.

extremely unusual feature for a fault of this size, although more typical of near-surface non-volcanic landslide basal detachments.

Stop 9.5 The San Andrés fault exposures at the road south of Tiñor   Past Valverde the TF-912 road runs up hill to the village of Tiñor. At the junction, take the old road to the left, around the explosion crater of La Caldereta. This old road passes through the village of Tiñor at the head of the canyon formed by the Barranco de Tiñor, where it cuts through the San Andrés fault scarp. South of the village the road follows the fault scarp with near-continuous exposures of flat-lying basaltic lavas and lapilli beds in the roadcut. These are of Jaramillo-age lavas (dated at 1.0 million years old) in the footwall sequence. The fault rocks themselves have been eroded away at this point.

At km-15, the exposures of lavas are replaced by steeply southwest-dipping rockfaces formed by weathered but distinctly grooved fault rocks. Stop at the side of the road. There is very little traffic, since this road has been replaced by a much larger highway farther to the west, and it now serves only Tiñor village and a few farms. It is therefore possible to use one lane for parking along straight stretches of road, as at this location.

The fault rocks are not as well preserved at this point as at stop 9.4, in part because of deep weathering but also because, having formed much closer to the palaeo-surface (note that this exposure is some 300 m higher than the Barranco de Tiñor exposures), they are less strongly indurated. A thin layer of cohesive breccia is present, but there is no evidence for the presence of pseudotachylyte. Many blocks projected through the main slip surface and were abraded into wedge-shape forms like roches moutonées; these indicate the slip sense (down to the east, as indicated by the inferred offset). Poorly preserved grooves and larger-scale undulations on the fault surface indicate the same direction of slip (i.e. down dip) as observed at stop 9.4. As in the Barranco de Tiñor, there is no evidence for polyphase slip in these fault-rock exposures. Along the road, the fault surface has eroded away in several places and a 1–2 m-thick zone of incohesive breccia is visible behind it.

To the east and south the fault scarp drops away to an area of relatively flat ground at the head of some barrancos. Mostly undeformed lava sequences are exposed in these barrancos, with only minor faults and joint systems exposed. To the south, however, beyond a farm complex, a north-east–southwest-trending ridge with a steep northwest face may have been produced by development of a significant antithetic fault. In general it appears that the complexity of the San Andrés fault system increases towards the south, but evidence for this is mostly obscured by younger, Rift lavas that have draped the fault scarp.

257

Itineraries in the El Golfo volcano

These itineraries will examine the rocks of the second (El Golfo) volcanic edifice, the Rift-series volcanic rocks (especially those filling the El Golfo embayment) and evidence for the formation, modification and age(s) of the El Golfo collapse structure. A particular problem with El Golfo is that onshore evidence points to an old (100 000–130 000 years?) age for the onshore structure, whereas it has been correlated with a much younger (15 000–9000 years) debris avalanche and megaturbidite off shore.

The eastern rim of the El Golfo collapse escarpment
Stop 9.6 Mirador Ermita de La Peña Begin from the village of San Andrés.[*] West of San Andrés, turn right onto the road to Mirador de La Peña. Hills to the east of the road in the northern part of the route are formed of flat-lying lavas of the Tiñor edifice; the line of hills marks the position of the buried Tiñor collapse scar.

Mirador de La Peña (not the modern mirador with a restaurant, but the old mirador located to the south at the Ermita de la Virgen de La Peña) overlooks the El Golfo embayment from its northern end. The cliffs at this point are about 700 m high and rise to over 1.2 km high to the south (Fig. 9.8a). The far side of the embayment at Punta Arenas Blancas is about 15 km away and the most southeasterly parts of the embayment cliff behind Frontera and Tigaday are 10 km behind the mouth of the embayment at sea level.

This spectacular viewpoint (best for photographs early in the morning) will be the first view of the El Golfo embayment[†] and an excellent place to begin a review of the stratigraphy of the pre- and post-collapse sequences and an overview of the structure of the El Golfo volcano.

Embayment-filling sequences
The elements of the embayment-filling sequence that are visible from this viewpoint are as follows:
- The young volcanic sequence, consisting of basaltic lavas that make up the platform between the foot of the cliffs and screes and the coastline (Fig. 9.8b). These lavas were partly erupted from vents on the floor of the embayment, but mostly from vents of the western rift zone, which breached the wall of the embayment and occur on its rim, including Montaña Colorada and the very large scoria-spatter cone of Tangana-

---

[*] The San Andrés fault passes beneath this village. . . even though it is concealed beneath Rift lavas, the similarity of its name to that of the famous San Andreas fault in California is too close to go unremarked.
[†] Actually a redundancy, since "golfo" is Spanish for "embayment".

Figure 9.8 **(a)** View of the 1 km-high cliff of the El Golfo gravitational collapse. **(b)** General view of the El Golfo collapse embayment and the town of Frontera from Malpaso (stop 9.10), the highest elevation of El Hierro.

soga to the southwest (Fig. 9.2). A few lava flows are intercalated with screes in the deepest part of the inner embayment east of Frontera; the vents for these form part of the north-northwest-trending rift zone. The oldest dated lavas of this sequence are not found on the surface but in boreholes north of Tigaday (2–5 in Fig. 9.3); these are $12\,000\pm7000$, $15\,000\pm2000$ and $21\,000\pm5000$ years old, in correct stratigraphic order. The oldest dated lava is not the deepest in the borehole but is close to the foot of the lava sequence.

- Screes and alluvial fans at the foot of the embayment cliffs, which occur above and intercalated with lavas. Some of these are young and still active, whereas an earlier generation of screes occurs only as remnants perched on the cliffs (Fig. 9.9a).

Taken in isolation, the radiometric ages of the screes indicate merely that the embayment is older than $21\,000\pm5000$ years. However, in many of the boreholes drilled during the course of water exploration on the floor of the embayment, a broad horizontal marine abrasion platform, up to 2 km wide, was found beneath the young lava sequence (see Fig. 9.3). This platform is near present sea level and extends from close to the back wall of the embayment between Sabinosa and north of Fuga de Gorreta, to about 1 km from the present coastline. On top of this platform, beneath the lavas, are deposits of aeolian sands, inferred to form a subaerial dunefield. The abrasion platform cuts into El Golfo series volcanic rocks and, east of Frontera, polymict breccias above and below El Golfo series lavas. These breccias may represent elements of the El Golfo and Tiñor debris avalanches, respectively, or of extensive post-collapse scree breccias. The formation of this abrasion platform indicates a long period of post-collapse erosion, the duration of which can possibly be evaluated by consideration of Quaternary sea-level variations.

The lack of pillow basalts or hyaloclastites above present sea level (see stop 9.2) in the Rift-series volcanic sequence, which is entirely subaerial, implies that there has been no recent uplift of El Hierro. Therefore, the abrasion platform must have formed when sea level was close to or above its present level. The last time this was so was about 100 000–130 000 years ago during the most recent interglacial (Fig. 9.9b). The subsequent fall in sea level would have exposed the abrasion platform, consistent with the development of a subaerial aeolian dunefield. This argument therefore implies that the El Golfo collapse took place at least 100 000 years ago and was followed by a 70 000-year-long period of volcanic repose before eruptive activity resumed in the embayment.

A maximum age for the embayment is in principle provided by the $133\,000\pm4000$ years age of the topmost lava in the cliff-forming sequence at Ermita de La Peña (stop 9.6, below). However, an interesting general

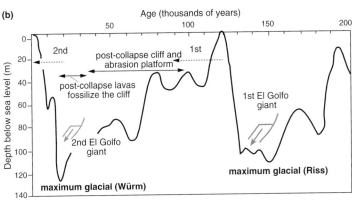

Figure 9.9 **(a)** Aerial view of the El Golfo collapse escarpment (stop 9.13), with the multiple generations of screes and coastal cliffs. **(b)** Sketch illustrating the possible relation of the collapse (or collapses) of the El Golfo volcano to the most recent two maximum glacials (stop 9.14).

problem in dating collapse structures is provided by the fact that cliff retreat to form the abrasion platform has undoubtedly taken place. Are these flows at the top of the present-day cliff in fact post-collapse flows that originally draped the collapse scar but which were subsequently eroded? It seems very probable that the El Golfo collapse postdates the trachytic rocks (c.176 000 years ago) erupted from the central vent of the El Golfo

261

edifice. However, both are maximum ages and it is also important to determine a minimum age of the collapse. Evidence for this will be discussed at stop 9.14.

A final feature of the embayment is the deep cusp or inner embayment to the east of Frontera. The origin of this feature is uncertain. It may reflect an original feature of the collapse structure, a later secondary collapse, or localized more-intense post-collapse erosion. It is possible that this embayment has been eroded along a westward extension of the sidewall fault of the San Andres aborted lateral collapse (p. 252), as occurred with the Las Playas barranco to the east; some palaeomagnetic evidence exists to support the existence of such an extension of the fault.

Collapse sequences before El Golfo
The same main stratigraphic elements (El Golfo and Rift series) seen in the Ermita de La Peña cliff section can also be seen in the central and western parts of the collapse escarpment. A north-northeast-trending swarm is prominent in the cliffs of Fuga de Gorreta to the northeast; the south-southeast swarm is exposed in a window high in the cliff (past km-27 on the road descending into El Golfo, stop 9.12 below); the final westerly swarm is best seen in the cliffs near Sabinosa to the west (stop 9.11).

Stop 9.7 Descent of the El Golfo embayment cliff via the Ermita de La Peña path
Plan in advance for a car to wait near the junction of the path with the road.

From the mirador, proceed via the clifftop road to the Ermita de La Peña. Continued rockfalls from the cliff have caused collapse of the old path down the cliff, and the new path starts from a small mirador a few tens of metres south of the Ermita. The descent of the cliff takes about 2.5–3.0 hours including stops and is vertiginous in places, although the path is wide.

The path provides an almost continuous section through this part of the El Golfo embayment cliff, and therefore it has been used extensively in palaeomagnetic and geochronological studies of the El Golfo edifice. A section through the cliff with radiometric sample locations and the main elements of the stratigraphy are shown in Figure 9.3.

The uppermost, Rift lavas form the sequence down to 620 m above sea level (dated sample at base of Rift series, 650 m above sea level, $158\,000\pm4000$ years) and are alkali basalts. Beneath the lowest lava, and separating it from the underlying trachytes, is a baked unconformity with thin discontinuous epiclastic breccia lenses.

Around 600 m above sea level (dated sample: $176\,000\pm3000$ years, at 585 m above sea level, in the middle of a trachyte unit) are the trachytic lavas that form the top of the El Golfo volcano in the eastern part of the embayment (Figs 9.2, 9.3). Three main lava flow units are present, with intercalated block-and-ash pyroclastic units. This assemblage implies the presence of a significant shallow magma reservoir at this stage of development of the El Golfo edifice, consistent with the structural evidence for the presence of an El Golfo central vent complex in the Frontera area (within the present embayment).

Below the trachytes is a sequence of alkali basaltic to intermediate lavas (530–400 m above sea level): the older members of the sequence are the least evolved. Rare lapilli and scoria layers occur between the lava flows, but major scoria cone units and dykes are not present along the path. A dated lava flow (261 000 ± 6000 years) occurs at about 500 m.

Stop 9.8 Mirador de Jinama   Mirador de Jinama is another spectacular viewpoint, best appreciated early in the morning. It is located at the highest vertical section of the collapse escarpment (1000 m); this is comparable in size and steepness to the famous northern cliff of Molokai (Hawaiian Islands), also a giant collapse escarpment.

From the mirador a track to the north follows along 500 m of the rim of the escarpment to Montaña Izique. Open fissures – precursors of future rockfalls – can be observed along the rim of the escarpment.

Stop 9.9 Descent of the El Golfo embayment cliff via the Mirador de Jinama path   Plan in advance for a car to wait near the junction of the path with the road.

From the Mirador de Jinama a 3.5 km path (similar to the Ermita de La Peña path) heads down the cliff to Frontera. The first part of the trail, to about 900 m, descends on basaltic flows of the Rift series. Farther down, two Strombolian cones appear attached to the cliff. Lava flows from these vents cascade down the cliff to the coast. The old trail took advantage of these flows to descend to Frontera. Sub-horizontal lavas densely intruded by dykes of the El Golfo volcano can be seen in gaps between the cascading Rift-series flows. The last stretch (from about 750 m down to Frontera) follows the gentler slopes of recent lavas and scree deposits.

The central and western sectors of the rim of the El Golfo collapse escarpment
From the Mirador de Jinama return to the main road, which runs close to the escarpment. The road cuts lapilli weathered to form a fertile soil. The abundant quarries were opened to extract this soil for farms at El Golfo. Along the road, the rim can be reached at several places, but the best is the Mirador del Golfo, just at the signpost of Hoya de Fileba (1300 m). This Hoya (a name for a large crater) is a hydrovolcanic explosion crater, in the rim of the escarpment. A short path leads to a magnificent view of El Golfo.

At the road junction, turn left and follow the road and then the dirt track to the easternmost edge of the escarpment.

Stop 9.10 Malpaso From the main track take a right turn signposted Malpaso, the highest point of El Hierro, a vantage point to take photographs of the east (best in the afternoon) and west (best in the morning) parts of the embayment.

About 100 m to the west of the main antennae there is an excellent view of the Tanganasoga volcano, the largest eruptive vent of the Rift series, with lavas cascading to fill the collapse embayment. The volcano is mantled by patches of pale trachytic ashes. These incorporate carbonized organic remains, dated at 6700 years old.

Stop 9.11 Overview of the El Golfo embayment from Mirador de El Rincón and Mirador de Bascos Return to the main track and proceed to the west to Mirador de El Rincón and Mirador de Bascos, the westernmost edge of the escarpment. Here there are excellent views of the entire embayment, especially in the afternoon (the opposite view of stop 9.6). The cliff facing you (behind Frontera) reveals a dense group of dykes emplaced beneath the northeast rift and exposed by the collapse.

The El Golfo collapse embayment

The eastern cliff

Stop 9.12 The road descending to Frontera As the road descends the cliff towards Frontera on recent (Rift series) lavas, a window past km-27 shows lavas of the El Golfo volcano, cut by several dykes. At km-30 there is an excellent mirador to observe the Fuga de Gorreta cliff (the view of Fig. 9.8b). Another fine closer view of this cliff can be had from the belltower of the chapel of Frontera.

Stop 9.13 The Fuga de Gorreta and rockfall scar From Tigaday, proceed north on the road to Punta Grande. About 3 km north of Tigaday, and just after passing under power lines, park on the right of the road and view the Fuga de Gorreta cliff to the northeast.

This cliff is about 1 km high and is mostly formed of El Golfo volcano lavas, cut by many dykes. A very fresh wedge-shape rockfall scar, some 600 m high and 200 m wide at the top, cuts the cliff at the western end of Fuga de Gorreta. The vertically elongated wedge-shape geometry may have in part been controlled by the dykes in the cliff. This rockfall occurred in 1910 and the talus cone that it formed buried the small village of Guinea at the foot of the cliff (now reconstructed as a museum). The total volume of the rockfall scar, which has enlarged since the initial event, is of the order of 1 million m^3 (approximately 0.0005% of the volume of a typical giant lateral collapse). Nevertheless, the repetition of many such events can considerably enlarge collapse scars over geological time.

Stop 9.14 Multiple generations of screes and coastal cliffs development and the age of the collapse Continue northeastwards along the road, stop 200 m or so beyond reconstructed houses at the foot of the 1910 rockfall at the junction with the track running southeastwards along foot of cliff. Walk 100 m or so along track to view the lower part of the cliff.

The west-facing cliff north of Fuga de Gorreta provides the best surface exposures of evidence for the marine abrasion platform found via the boreholes (see Fig. 9.9a). The lowest 50 m or so of the cliff is vertical and relatively fresh (C2), whereas higher parts of the cliff (C1) are steeply inclined, more intensely weathered and vegetated (L1). Furthermore, perched on this higher part of the cliff are the truncated remains of screes that pre-date the formation of the lower cliff (S1). The latter is a coastal cliff that formed at the back of the marine abrasion platform, the abrasion having removed the earlier-formed screes (first-generation screes of Fig. 9.3) that had developed at an earlier stage of retreat of the collapse scar.

The embayment-filling lava flows (L2) are exposed almost to the foot of the cliff with only very limited development of later screes (S2), suggesting that very little further erosion took place after the abrasion platform emerged above sea level. This succession of geological events makes it very difficult to accept a young (*c.*15 000 years) age for the subaerial El Golfo collapse, as discussed below (stop 9.15).

Stop 9.15 Punta Grande and los Roques de Salmor Continue to the end of the road, to where it descends to the coast at Punta Grande, the northernmost end of the embayment-filling lava sequence as presently preserved. To the north the El Golfo cliff is still being eroded by the sea, and exposures of the pre-collapse sequences of rocks are particularly good. Coastal bathymetric data suggest that the cliff in this area has retreated 0.5–1 km in the recent past, leaving a shallow abrasion platform. The problem of whether the topmost lavas in the cliff are pre- or post-collapse is therefore particularly acute in this area; possible cliff-draping lavas are visible at the top of the cliff near the truncated scoria cone north of the Mirador de La Peña (stop 9.6).

The cliff-forming sequence shows strong lateral variations in this area (Fig. 9.10a). The trachytes at the top of the El Golfo series are truncated in places and appear to fill channels elsewhere, especially at the northern point of the cliff where they drop to sea level. The Roques de Salmor (Fig. 9.10b) are almost entirely formed of these trachyte lavas.

The very rapid lateral variations in the cliff-forming sequence and the steep northward dips of many units suggest that these outcrops lie close to the pre-collapse palaeo-coastline and that no large (more than a few hundred metres wide) abrasion platform has developed by retreat of the coastline since the El Golfo collapse. The contrast with the rapid development of an abrasion platform within the embayment probably reflects exposure of the soft and easily eroded pyroclastic rocks of the lower El Golfo series by the collapse itself.

The recent coastal cliffs cut into the embayment-fill sequence reveal lava flows, small scoria and spatter vents, but no evidence of hydrovolcanism, not even distal hydrovolcanic ash layers between lavas. There is no evidence in these exposures of proximity to the coastline at the time of formation of these lavas, suggesting that at the time of its maximum development the embayment-filling lava platform extended much farther off shore (1–2 km?). However, the available bathymetry indicates a very narrow coastal abrasion platform, 100–200 m wide, beyond which a very steep submarine slope is present. How has the distal part of the lava platform been removed? We suggest that the solution to this problem may also

Figure 9.10 **(a)** Main geological units at the northern edge of the El Golfo collapse escarpment (stop 9.15). **(b, opposite)** Roques de Salmor from the sea. These are eroded remnants of the terminal trachytic eruptions of the El Golfo volcano.

explain the discrepancy between the ages of El Golfo indicated by the onshore and offshore evidence. Boreholes between the coastline and the edge of the old marine abrasion platform, more than 1 km inland, cross only subaerial lavas around the present sea level. The post-collapse embayment-filling lavas may have extended out beyond the abrasion platform onto the very steep slope formed by the submarine part of the older (later than 133 000 years ago) collapse structure, forming a thick and unstable lava-deltaic sequence that extended well out beyond the present coastline in the central part of the embayment. Collapse of this sequence, and perhaps also of some of the underlying rocks, could have produced a second debris avalanche, blanketing the first, and an associated turbidite at 10 000–15 000 years ago, as observed in the offshore sequences. The discrepancy may therefore be resolved by postulating two collapses rather than one, each of them occurring just after the lowest sea level of the most recent two maximum glacials (Fig. 9.9b), or during sea-level rise in the early part of the succeeding interglacial.

The western cliff

Stop 9.16 The Sabinosa window   Rejoin the coastal road and continue south-west on tracks along the coast to Sabinosa to the Pozo de La Salud Hotel to view a window into El Golfo series rocks southwest and west of Sabinosa.

Most of the southern wall of the El Golfo embayment is draped with young Rift-series lava flows and volcanic vents, one of which (Tangana-soga) is very large (stop 9.10). West of the village of Sabinosa, a small steep-walled embayment, possibly representing the scar left by a recent moderate-sized (2–4 km^3) slope failure, exposes rocks of the El Golfo edifice. The cliff is composed of El Golfo series rocks containing many west to west-northwest-trending dykes, representing the western rift zone of the El Golfo volcano.

Stop 9.17 The Playa Blanca cliff and platform-forming eruptions   Proceed westwards along coastal track, rounding the promontory of rock at the western end of the El Golfo embayment. Park just outside the embayment to view at the foot of the palaeo-cliff a group of recent vents that have fed platform-forming lavas.

The development of a cliff-forming unit (El Golfo edifice rocks, plus some older Rift-series lavas) and a distinct platform- and scree-forming unit, as seen in the Cumbre Vieja of La Palma, is evident at this location. However, the vents feeding the platform-forming lavas are mainly located in or at the foot of the cliff face. This section of cliff is where part of the western rift zone reaches the coast (Fig. 9.2). The concentration of vents in the cliff implies that the dykes in the rift zone propagate laterally beneath the surface of the rift and emerge where the surface drops away, as at the cliff. Does this imply that dykes are capable of propagating laterally just a few hundred metres below the surface or do they migrate upwards beneath topographic lows?

The coastal topography at this location is in marked contrast to that at Sabinosa, a short distance to the east but inside the embayment. The transition from steep coastal terrain and cliffs, cut into young lavas at Sabinosa, to the more usual cliff and platform morphology at this location commonly occurs at the boundaries of collapse structures, even when the collapse embayment has been mostly filled by younger lavas, as at the La Orotava Valley in Tenerife. Coastal morphology may provide an important clue to the presence of such buried structures.

Stop 9.18 The tuff ring of Hoya del Verodal (2.11) Continue westwards along the coastal track, passing through young and fresh lava flows; park on the left about 300 m east of yellow palagonite tuff ring outcropping in the cliff. An abandoned track climbs to and old galería, excavated in the tuffs, which allows observation of the internal structure of the tuff ring.

This outcrop is one half of a large Surtseyan tuff-ring exposure (Fig. 9.11a), also exposed in the cliff to the southwest, consistent with the north-easterly dip of beds in the tuff outcrop, the crater being at least 1.1 km in diameter. Subaerial lavas of the El Golfo volcano cover the tuff ring.

The presence of this tuff ring at (more or less) present sea level again demonstrates the stability of El Hierro relative to sea level, at least during the past 250 000 years (see also stop 9.2).

Young lava flows and the problem of the supposed 1793 eruption
Continue westwards on the coast road to where it begins to climb up the cliff, passing very young lava flows and spatter vents. At the second bend, turn off onto a rough track and park immediately, at the southwestern limit of the very young lavas.

These lavas have been attributed to an eruption thought to have taken place in 1793. However, examination of the original eye-witness account shows that it contains clear references to a swarm of earthquakes, but none at all to the likely manifestations (fire fountains, lava flows, explosions and steam clouds from where the lava entered the sea) of the eruption that produced these lavas, which would have been clearly visible both to the inhabitants of the island and to observers on ships and fishing boats off shore. It therefore seems likely that these very fresh lavas were produced in a sub-historic eruption and have been well preserved because of the aridity of this part of the island. The earthquake swarm of 1793 may have been associated with the intrusion of a swarm of dykes or a submarine eruption well off shore.

The cliffs behind the coastal lava platform contain the other half of the Surtseyan tuff ring and a bisected Strombolian cone perched on the top of the cliff with its feeder dyke exposed.

Itineraries in the Rifts

The northwestern rift
The best procedure to observe the northwest rift is to follow the road up hill from stop 9.18 to Faro de Orchilla.

Figure 9.11 **(a)** Western half of the hydrovolcanic tuff ring of Hoya del Verodal (stop 9.18) at the west end of the island. **(b)** The Lomo Negro vent and lavas, a prehistoric eruption incorrectly associated with a seismic crisis (probably a submarine eruption) on the island in 1793.

Figure 9.12 **(a)** Aerial view of the Montaña Las Calcosas volcanic group and the Orchilla lighthouse (stop 9.19). **(b, overleaf)** Aerial view of El Julan collapse embayment (stops 9.19 and 9.21), filled with lavas from the northeast rift.

Stop 9.19 The lighthouse or Faro de Orchilla From the descending road, impressive views can be observed of recent groups of vents and lava flows (such as the Montaña Las Calcosas and Montaña de Orchilla, shown in Fig. 9.12a) of the northwestern rift. To the east can be seen the El Julan collapse embayment, entirely filled with lavas from this northwestern branch of the Rift volcanism (Fig. 9.12b). Spectacular views of the El Julan embayment can be obtained at many places along the road signposted Hoya del Morcillo–El Julan, cut into the Rift lavas cascading down to fill the embayment.

Past Hoya del Morcillo recreation area, join the main road to El Pinar and La Restinga, along the southern branch of the Rift system.

The southern rift and El Julan

Stop 9.20 Mirador de Tajanara Just before the village of El Pinar, turn right and right again to the mirador on top of Montaña Tajanara. This mirador provides a view over the scoria cones of the south-southeast-trending rift zone of El Hierro. This rift zone has many scoria cones and lava flows of recent appearance (especially at its admittedly arid southern end), which form extensive coastal lava platforms (Fig. 9.2).

Stop 9.21 Views of the El Julan lava field and embayment Return to the main road towards La Restinga, but turn off down the lane to Tacorón, crossing young pahoehoe lavas of the south rift zone. Continue down to the coast at Cala de Tacorón to view the El Julan embayment.

The El Julan embayment dominates the topography of the southwest-ern side of El Hierro (Fig. 9.12b). Bathymetric data also indicate that it extends well off shore. It was first identified as a lateral collapse structure on the basis of seafloor-imaging sonar data that revealed the presence of a large debris-avalanche deposit passing under the Saharan debris flow to the south (also indicative of the relative antiquity of the debris avalanche). There are no exposures of the embayment wall or the pre-collapse series of rocks (probably El Golfo edifice rocks) on shore, the embayment having been entirely buried by post-collapse lavas. At either end of the em-bayment, these include relatively young lavas that pass laterally into platform-forming lavas, but, in the centre, cliffs over 100 m high are present with no evidence of a coastal lava platform. Lavas at the top and foot of this cliff have been dated at 41 000 and 31 000 years old, respectively (see 3 in Fig. 9.3). However, this gives only a minimum age for the collapse, which is much older.

Although the lavas at either end originate respectively from the south-southeastern rift zone and the western rift zone beyond the limit of the El Golfo embayment (Fig. 9.2), the lavas in the centre would have origi-nated from vents that are now at the crest of the El Golfo escarpment or have been removed during its formation. Could the cessation of eruption of lava flows to the south from this section of the western rift be related to the El Golfo collapse, implying a large time interval between the El Julan and El Golfo collapses? Alternatively, it may be related to the subsequent enlargement of the El Golfo embayment by erosion.

On returning to the main road, take a track descending to a large green-house built on top of the El Julan volcano lava field. Walk along the lava field to the east to observe one of the best examples in the Canaries of pahoehoe and aa lava flow morphologies. Spectacular structures can be observed and photographed. Pahoehoe lavas cascade from **spatter cones** (Fig. 9.13a), or extrude from **tumuli** and **pressure ridges** (Fig. 9.13b), to form "ropes" and "toes" (Fig. 9.13c). Some of the tumuli and pressure ridges have triangular wedges of fluted lava protruding from the medial cracks.

Figure 9.13 Typical features of a pahoehoe lava field, coastal lava platform of El Julan volcano (stop 9.21), near La Restinga (south end of the island): **(a)** spatter cone and cascades of pahoehoe lavas; **(b, opposite)** pressure ridge refilled with wedges of fluted lava protruding from the medial cracks; **(c, p. 276)** curious forms of pahoehoe "ropes" and "toes".

Glossary

Spanish words for topographic and other features appear in *italics*.

aa The Hawaiian name for a type of lava in which the first-formed solid crust is broken up by continued rapid flow; this continues until the lava flow consists of a mass of irregular congealed fragments with a rough top surface, around a core of solidified lava.

accretionary lava balls A rounded mass up to several metres in diameter, formed by moulding lava around a core when flowing down slope (similar to a snowball).

alkali basalt A volcanic rock with a magnesium- and iron-rich silica-poor composition (typically one that has ascended from its mantle source without much modification in its composition by crystallization), but also with a relatively high content of alkali elements (Na, K) attributable to having been formed by only little partial melting in its mantle source.

almagre From the Arab *al-magre* (red clay). Originally referred to a red pigment obtained from soils baked and oxidized by heat from lava flows. True almagres or baked soils imply a long interval between flows in a lava sequence; the term should not be used for baked lapilli beds, because these do not imply such an interval.

barranco A gulley, valley or canyon produced by rivers or flash floods; the word covers features from 10 m to over 1 km deep.

basanite Like an alkali basalt, but even more alkali-element rich and silica poor; produced by even less partial melting in the mantle.

block-and-ash deposit Deposit from an incandescent cloud or glowing avalanche, formed by a mixture of fragmented volcanic rock, ash and gas released at high speed in an explosive eruption. In La Palma these deposits are formed by collapses of juvenile incandescent phonolite lava domes.

breccias A coarse-grain clastic rock composed of cemented angular broken rock fragments. May be sedimentary, volcanic, or produced by intrusion or faulting.

brunhes See **geomagnetic polarity**.

caliche A calcium or magnesium carbonate-rich soil produced by dissolution and re-precipitation of carbonates, either from the underlying rocks or from sediments rich in limestone fragments or marine shell debris.

carbonatite The only terrestrial igneous rock type not primarily composed of silicates, carbonatite is composed instead of carbonate minerals with lesser amounts of various Ca, Ti and Fe oxides.

cataclasite Fragmented rocks produced by shattering and crushing along a moving fault; may be coarse, loose breccias or fine-grain cohesive rocks, depending on the intensity of the crushing.

cinder cones A conical hill formed by accumulation of juvenile pyroclastic fragments falling to the ground in an essentially solid state. Cinder and **tephra** are frequently used as general terms for all pyroclastics of a volcano.

collapse calderas Large craters produced by subsidence or collapse of the crater floor as an underlying magma chamber is emptied in a large eruption.

cone sheet A curved intrusion, formed in a manner similar to a **dyke**, but emplaced in a radial stress field around a large intrusion, resulting in a shape that forms part of an inverted cone with its point within the source intrusion. Usually forms only a small part of the cone, whose shape can only be discerned from many cone sheets forming a cone-sheet swarm.

conglomerate See **breccia.**

coulée A short thick lava flow, formed by an eruption of viscous magma, with characteristic steep sides and a top with transverse ridges that are convex in the direction of flow.

cryptodome A large blob of viscous magma that expands just beneath the surface, rather than erupting, and lifts the ground surface up by tens or hundreds of metres. It may break through the surface as it enlarges.

cumbre The crest of a ridge, or the plateau between valleys.

cumulate A coarse-grain igneous rock composed of crystals of one or more minerals that have separated out from a magma during crystallization; its composition is governed by the minerals present and not directly by the composition of the parent magma.

debris-avalanche deposits Deposits produced by the collapse and disintegration of the flank of a volcano, producing a mass of rock debris that moves down slope as a fast-moving giant avalanche that, depending on its size, can travel for tens or even hundreds of kilometres. Characteristically composed of all sizes of fragments, from mud to very large blocks; in the largest debris avalanches these blocks can be tens of kilometres across.

differentiated A magma lacking the original composition of the partial melt of crust or mantle from which it formed, but has instead been modified, typically by cooling and partial crystallization of the magma, followed by separation of the denser crystals to leave a magma of a new composition. Since the separated minerals are typically (but not always) richer in Fe and Mg, and poorer in Si and alkali elements, than the remaining melts, differentiated

magmas are generally richer in the latter elements; see **felsic**.

distal The area far from the vent of an eruption or a sediment source; pyroclastic rocks or sediments deposited in such an area, typically of finer grain and as thinner units than the corresponding deposits in the **proximal** areas near the vent or source.

dorsal Term used in the Canaries for a topographic ridge (typically formed by eruptions along a volcanic **rift zone**).

dolerite A medium-grain (crystal sizes 0.5–2 mm typical) igneous rock of basaltic composition, often (but *not* necessarily) forming a small intrusion such as a dyke.

dome A steep-sided, flat- or convex-top body of igneous rock, formed either by extrusion of highly viscous lava on the surface or by emplacement of a similar magma just below the surface. Domes of the latter type are sometimes referred to as **cryptodomes**.

dyke An igneous intrusion formed by cracking of the surrounding rocks as magma ascends to the surface or moves laterally; the resulting fracture fills with pressurized magma and is self-propagating until the magma solidifies, producing a thin but laterally and vertically extensive sheet of solidified igneous rock that cuts across pre-existing structures such as bedding or earlier intrusions. Dykes often occur in large numbers, forming a **dyke swarm**, and are typically near vertical, although strictly it refers to any such intrusion that cross-cuts pre-existing sedimentary bedding or other geological contacts.

epiclastic A sedimentary rock composed of pyroclastic or other fragments of volcanic rock that have been eroded and re-deposited with little or no compositional modification (by weathering or addition of non-volcanic rock fragments).

extrusive An igneous rock emplaced as a magma at the surface of the Earth.

fan delta A gently sloping alluvial deposit produced where a stream flows into a lowland.

fault gouge A soft, fine-grain, often

clay-rich rock produced by a combination of physical crushing and chemical alteration (by reaction with groundwater) of rocks along an active fault.

felsic A very general term for igneous rocks with high contents of Si, Al, K, Na (elements that form mainly pale minerals) and lower contents of Fe and Mg (elements that form dark or **mafic** minerals). Particularly useful as a general term for the alkali-rich, Fe- and Mg-poor rocks of the Canaries and other oceanic island volcanoes (**trachytes** and **phonolites**); in these volcanoes the terms "acid" and "silicic", corresponding to the silicic calc-alkaline volcanoes (e.g. most volcanoes of the Pacific coast or "ring of fire"), should be avoided.

fiammé Flattened and streaked out but still recognizable pumice clasts within an **ignimbrite**. The deformation occurs just after deposition while the rock is still hot and the glass making up the bulk of the pumice can deform plastically as the rock compacts and flows under its own weight.

flow banding Compositional or textural banding in a lava or ignimbrite, produced by intense streaking out of heterogeneities in the rock as it slowly flows while still hot and plastic.

fumarolic Alteration and recrystallization produced by reaction with hot volcanic gases percolating through the rocks thus affected.

gabbro A coarsely crystalline igneous rock mainly composed of the minerals plagioclase feldspar and pyroxene, often with olivine as well. The compositional equivalent of fine-grain igneous rocks such as basalt and basanite.

galerías Horizontal tunnels (~2 m in diameter), typically 0.5 km to 5 km long and totalling about 3000 km in length, excavated to mine groundwater in the Canaries, mostly in Tenerife and La Palma. They are an extremely useful way of gaining access to the internal structures of the island volcanoes for observation and sampling.

geomagnetic polarity Changes in the Earth's magnetic field between normal polarity and reversed polarity occur at intervals and form the basis of magnetostratigraphy. Radiometrically dated reversals define the geomagnetic polarity timescale. This is composed of named magnetic polarity epochs. The youngest (and continuing) is the **Brunhes** normal-polarity epoch (0.78 million years ago to present); the preceding **Matuyama** reversed polarity epoch lasted from 1.8 to 0.78 million years ago. The polarity epochs of predominantly or entirely one polarity have polarity events, the shortest polarity-chronological unit, as the **Jaramillo event** (1.053 to 0.986 million years ago) of normal polarity in the Matuyama epoch. The magnetic reversal event between two epochs is known as a boundary and known by the combination of epoch names, for example "Matuyama/Brunhes".

glacial maximum The time of the greatest extent of a glaciation. The most recent **glacial maximum** was reached between 14 000 and 22 000 years ago. In the recent geological past, each glacial maximum corresponds to a marine **lowstand**.

graben A long narrow trough produced by two parallel inward-facing faults. Produced by extension of the crust (at the large scale) or by local small-scale extension, for example above a dyke whose tip is just below the surface.

greenschist-facies metamorphism Metamorphic facies corresponding to temperatures in the range 300–500°C, transforming the primary minerals in the rocks to a set of metamorphic minerals (in particular albite, epidote, chlorite), characteristically white and green in colour.

highstand A period when sea level is high and relatively stable (e.g. during interglacials or other periods of warm climate).

Holocene The period of Earth history since the end of the most recent ice age, around 10 000 years ago.

hotspot Characterized by raised topography, high heat flows and volcanic activity, the surface expression of a **mantle plume**.

hyaloclastite A rock composed of fragments of altered volcanic glass, formed by the flowing of lava into water, leading to explosive mixing; see **hydrovolcanic**.

hydrothermal alteration   Alteration of rocks by circulating groundwaters heated by igneous activity.

hydrovolcanic   Explosive volcanic activity driven by the violent expansion of heated boiling groundwater that mixes with the magma near or at the surface, rather than by escape of gases dissolved in the magma at great pressure and released as it ascends to the surface from depth.

ignimbrites   A rock formed by the deposition and consolidation of hot fragments of disintegrated viscous lava and ash in explosive eruptions, at such a rate that the rock is still hot and undergoes variable amounts of compaction, welding and flow, as it accumulates producing characteristic features such as **fiammé** (see also **nuèe ardente**).

inclined sheet   An inclined sheet intrusion (see **dyke**), formed in an obliquely orientated stress field.

intrusion   A body of rock formed by emplacement of magma as a mass underground, typically by forcing apart the rocks into which the intrusion is emplaced, followed by solidification *in situ*. Intrusions are composed of **intrusive rocks**.

inversion of relief   Young lavas filling a river bed or valley are usually more resistant than the host rock and erosion cuts into the latter, leaving the former in relief.

Jaramillo event   See **geomagnetic polarity**.

K/Ar radiometric age   See **radiometric age**.

keratophyre   **Felsic** extrusive rocks commonly interbedded with submarine lavas and sediments.

lahar   An Indonesian word used to name a volcanic mudflow travelling down valley, generally formed by mixing of rain with ash, or when an eruption displaces a lake or melts part of an icecap (e.g. the Nevado del Ruiz mudflow, Colombia, 1985).

landslide deposits   Deposits that are typically dominated by rock fragments or soil, transported down slope and redeposited in landslides. Characteristically poorly sorted, often including very large blocks. See also **debris-avalanche deposits**.

lateral collapse   A large landslide that removes a major fraction of the flank of a volcano; may be triggered by intrusion of magma, hydrothermal weakening and groundwater pressurization, or other mechanism. Collapse volumes range from cubic kilometres to hundreds of cubic kilometres.

lava domes   Erupted masses of highly viscous lava that pile up around the vent rather than flowing away as lava flows; see **dome**.

lithic clasts   Fragments of pre-existing rocks eroded and included in a sedimentary rock, as opposed to reworked volcanic scoria or other debris, or bioclasts derived from living organisms such as shells.

lithic fragments   In pyroclastic rocks, pieces of pre-existing rocks that were incorporated into the fragmented magma; distinct from juvenile fragments or solidified pieces of the magma that was involved in the eruption.

lowstand   A period of low and uniform sea level, for example during the coldest part of a glaciation.

maar   A broad volcanic crater sunk below the level of the surrounding land surface, with at most only a low rim of volcanic deposits; produced by violent hydrovolcanic explosions.

mafic   Relating to iron- and magnesium-rich minerals, such as olivine and pyroxene, or to igneous rocks rich in these, such as basalts and gabbros.

mafic enclaves   Small elliptical or cauliflower-shape inclusions of mafic rock, interpreted as blobs of mafic magma which were chilled and quickly solidified when they mixed into lower-temperature, more felsic magmas before the latter solidified.

mantle plume   A concentrated upwelling in the convecting part of the Earth's mantle, in which hot peridotites rise rapidly from deep in the mantle and melt as they rise, to produce basalts and other magmas.

Matuyama See **geomagnetic polarity**.

Matuyama/Brunhes See **geomagnetic polarity**.

megacryst A general term for a much larger crystal in a finer-grain igneous rock, either formed by the initial stages of crystallization or incorporated into the magma from mantle rocks.

microgabbro A relatively fine-grain **gabbro**, with uniform crystal size between around 2 mm and 5 mm.

microsyenite A relatively fine-grain **syenite**, with uniform crystal size between around 2 mm and 5 mm.

Miocene The epoch of Earth history between 22 and 5.5 million years ago.

MORB Acronym for mid-ocean ridge basalt, produced by eruption at oceanic spreading centres where the tectonic plates are moving apart. Characteristically poor in Fe, alkali elements and trace elements compared to other types of basalt, because it is produced by melting of the mantle very close to the surface.

nephelinite An extremely alkali-rich, magnesium-rich and silica-poor volcanic rock.

nuée ardente See **block-and-ash**.

ocean-island tholeiite A type of basalt produced by relatively high degrees of partial melting at the most active **hotspots**, poorer in alkali elements and richer in silica than most such basalts.

oceanic lithosphere The cold rigid layer beneath the ocean basins, consisting of the thin oceanic crust and a layer of mantle underneath. This rigid layer, at most only 150 km or so thick, forms most of the tectonic plates that move around on the surface of the Earth.

ophiolite A slice of oceanic crust (and sometimes underlying mantle rocks) thrust up on top of adjacent continental crust as two plates converge, providing a section through the sequence of rocks (including **pillow basalts, dykes, gabbros**) that make up the oceanic crust.

orographic cloud Cloud formed by cooling and condensation in moist air currents rising on the windward side of mountains, yielding orographic rain.

outliers Isolated outcrops of younger sedimentary or volcanic rocks surrounded by older rocks, formed by erosion of a formerly continuous layer of rock.

pahoehoe The Hawaiian name for a type of lava with a smooth or wrinkled top surface, formed when a lava is fluid enough to flow without its surface crust fracturing (see **aa**); usually forms in near-horizontal landscapes where flow rates are low, but exceptionally fluid lavas form pahoehoe even on very steep slopes.

peralkaline An igneous rock with a high content of alkali elements (especially Na and K) in relation to its aluminium content, and therefore characterized by the presence of alkali-bearing amphiboles or pyroxene, or both, in addition to alkali feldspars.

peridotite The rock type that forms most of the upper mantle, consisting primarily of the mineral olivine (magnesium–iron silicate) with lesser amounts of pyroxene and one of several aluminium-bearing minerals (plagioclase, spinel or garnet), depending on the depth below the Earth's surface (and therefore depending on the pressure). Peridotites are the most important source of magmas in the mantle.

phenocrysts Relatively large, well shaped crystals in generally fine-grain igneous rocks, formed by slow partial crystallization of a magma at depth before it is intruded or extruded and solidifies rapidly, producing the finer-grain remainder of the rock.

phlogopite A magnesium-rich mica mineral, typically forming dark brown to red shiny flakes. Often found in extremely silica-poor alkali-rich rocks including carbonatites.

phonolite A fine-grain, felsic, differentiated igneous rock, relatively poor in silica but very rich in Na, K and Al, so mainly composed of alkali feldspars. Often also contains a feldspathoid mineral. Phonolites are easy to recognize in La Palma because they contain hauyne, a bright blue feldspathoid.

pillow lavas Interconnected lava tubes resembling pillows in cross section, formed in volcanic eruptions in subaqueous environments, especially in deep water where the pressure prevents **hydrovolcanic** explosions.

Pleistocene   The epoch of Earth history from about 1.8 million years ago to the end of the most recent ice age around 10 000 years ago. The Pleistocene/**Pliocene** boundary is generally established at 1.77 million years ago, the upper limit of the Olduvai normal **polarity** event in the **Matuyama** epoch.

Plinian   A very violent explosive volcanic eruption, producing an eruption column tens of kilometres high. Named after the Roman author Pliny the Younger, who described the first historically recorded Plinian eruption, that of Vesuvius in AD 79.

Plinian airfall   Pyroclastic deposits, characteristically very well sorted and extending over wide areas, produced by deposition from the eruption clouds formed in **Plinian** eruptions.

Pliocene   The epoch of Earth history from 5.5 million to 1.8 million years ago.

polymict   A breccia or other sediment composed of grains or clasts with a variety of compositions and from different sources.

porphyry   A rock, typically **felsic** in composition, with abundant and prominent phenocrysts.

pressure ridges   Ridges in the surface of a lava or other flow (such as a glacier) produced by uplift of the surface of the flow in response to an increase in pressure in the interior of the flow.

primitive   A rock or magma with a composition close to that when it was first produced by partial melting, modified only slightly if at all by **differentiation**.

proximal   Close to a vent or sediment source; see **distal**.

pseudotachylyte   A black glassy or very fine-grain rock along a fault, produced by frictional melting of the rocks on either side of the fault as they slid past each other.

pumice   A lump of frothy or sponge-like volcanic rock, typically **felsic** in composition, with a very low density, because it is mostly composed of originally gas-filled bubbles. Formed in

explosive (especially **Plinian**) eruptions by the violent expansion of gases coming out of solution in the magma as it decompresses.

pyroxenite   A coarse-grain intrusive igneous rock, composed entirely or almost entirely of crystals of one or more of the pyroxene group of minerals (Ca–Mg–Fe silicates).

radiometric age   A radiometric age of an igneous rock is derived from the ratio of abundances of a radioactive parent isotope and a daughter isotope produced by the radioactive decay of the parent: the older the rock, the more of the daughter isotope is present. The potassium/argon (K/Ar) parent–daughter pair is particularly useful in dating young igneous rocks.

rheomorphic ignimbrites   Ignimbrites hot on deposition to fuse and flow like viscous lava flows.

rift zones   Linear zones of concentrated extension, separated by zones of lesser deformation. This extension may involve zones of sub-parallel faults (as in the East African Rift Valley), or emplacement of many dykes. The former are sometimes called tectonic rift zones, the latter, volcanic rift zones; rift zones where both processes operate are common.

scoria   Fragments of frothy or vesicular (bubble-rich) basaltic volcanic rock; typically less bubble-rich than **pumice**.

screen   A thin layer of older rocks between two sub-parallel dykes or other intrusions emplaced side by side.

seamount   A large (500–5000 m-high) volcano that has grown on the ocean floor. Most seamounts never reach the surface, but some grow above sea level to form oceanic-island shield volcanoes; old volcanoes of this type sometimes subside below sea level to become flat-top seamounts or guyots.

shield volcano   A volcano, often very large, with gentle slopes that commonly decrease towards the summit. Usually composed mainly of lavas of basaltic or similar composition, and formed by many eruptions that build the volcano up over thousands or millions of years. Often has radial volcanic

rift zones from which most of the lavas are erupted.

sill A sheet-like intrusion (see dyke) emplaced along pre-existing sedimentary bedding (perhaps not horizontal).

spatter Scoria hot enough and fluid enough when it lands to spatter into deformed irregular masses; often ejected rapidly enough from a vent for the hot lumps of spatter to weld together, forming a steep-sided **spatter cone**.

stratovolcano A volcano formed by many eruptions, variously explosive or effusive, that build up a cone with a summit crater and slopes that steepen upwards.

surge deposits Pyroclastic deposits formed by deposition from fast-moving dense ash clouds surging outwards over the ground from a vent during a violently explosive (often hydrovolcanic) eruption. Characteristically of fine grain and with cross bedding and dune beds as a result of the high wind-speeds developed within the surge ash clouds.

syenite A coarse-grain igneous rock, equivalent in composition to **trachyte** or **phonolite**, and composed almost entirely of alkali feldspars.

tephrite A fine-grain differentiated rock produced by small amounts of crystallization from **basanite**; also silica-poor and alkali-rich, but with less Mg and Fe.

tholeiite A type of basalt, poor in alkali elements and relatively rich in silica. Produced by relatively high degrees of partial melting of a mantle source or melting, or both, of the mantle at shallow depths, so characteristic of only the most productive volcanic provinces such as the mid-ocean ridges and the hottest hotspots.

trachyte A fine-grain felsic igneous rock, rich in alkali elements but also in silica; intermediate in composition between **phonolites** and the most silica-rich felsic rocks, such as rhyolites. Trachyte lavas are viscous and are frequently involved in explosive eruptions or forming domes.

trap A topographical term, referring to the alternation of steep slopes and flats that develop on hillsides formed by alternating lava flow cores and weak flow tops and interflow sediments.

tsunami A very long-wavelength ocean wave, affecting the whole of the water column down to the floor of the ocean, produced by earthquakes, volcanic eruptions, landslides or even asteroid impacts into the ocean; capable of speeds comparable to those of a jet aircraft (600–800 km per hour).

tuff cone A volcanic cone composed mainly of fine ash and breccia produced by hydrovolcanic explosions, typically high sided but with a deep central crater often as deep as the cone is high (but see **maar**).

tuff ring A version of a tuff cone, but wider and lower; appears as a low volcanic cone with a very broad and shallow central crater surrounded by a low rim of hydrovolcanic ash.

tumuli Mound-like bulges in the surface of a lava flow produced by uplift of the surface in response to an increase in pressure in the core of the flow. for example in a lava tube.

turbidites Distinctive sand or silt beds in deepwater sedimentary sequences, produced by deposition from a sediment-rich current in which the sediment is maintained in suspension by turbulence.

volcaniclastic Any clastic rock containing a significant amount of volcanic material in any proportion and independently of origin or depositional environment.

xenolith A fragment of solid rock (but which may have been partially melted) incorporated into a magma as it ascends or is erupted. The term covers everything from fragments of the mantle brought up from just above the source of the magma, 100 km or more down, to pieces of pre-existing rocks exposed in the vent wall that fell into the magma as it broke the surface.

Further reading

General volcanology

Francis, P. W. 1993. *Volcanoes: a planetary perspective*. Oxford: Oxford University Press.
Scarth, A. 1994. *Volcanoes: an introduction*. London: UCL Press.
Sigurdsson, H. (ed.), 2000. *Encyclopedia of volcanoes*. London: Academic Press.
Tilling, R. I. 1999. *Volcanoes*. Denver, Colorado: United States Geological Survey (available online at http://pubs.usgs.gov/gip/volc/).
Tilling, R. I. & J. J. Dvorak 1993. Anatomy of a basaltic volcano. *Nature* **363**, 125–33.

General background

GENERAL INFORMATION
http://es.dir.yahoo.com/zonas_geograficas/paises/espana/
    comunidades_autonomas/canarias/
English-language version at:
http://dir. yahoo. com/Regional/Countries/Spain/
Rochford, N. "*Landscapes of...*" series: *Fuerteventura, Gran Canaria, Northern Tenerife, Southern Tenerife and La Gomera, La Palma* and *El Hierro*. London: Sunflower Guides. [These guides are frequently updated and reissued.]

HISTORY
Castellano, J. M. & F. Macías Martín 1993. *History of the Canary Islands*. La Laguna, Tenerife: Editorial Centro de Cultura Popular Canaria.
Torriani, L. 1592. *Descripción e historia del reino de las Islas Canarias* (Spanish translation by A. Cioranescu). Santa Cruz de Tenerife: Editorial Goya. 1978.

ARTICLES
Abdel-Monem, A., N. D. Watkins, P. W. Gast 1971. Potassium–argon ages, volcanic stratigraphy, and geomagnetic polarity history of the Canary Islands: Lanzarote, Fuerteventura, Gran Canaria and La Gomera. *American Journal of Science* **271**, 490–521.
Ablay, G., G. G. J. Ernst, J. Marti, R. S. J. Sparks 1995. The ~2 ka subplinian eruption of Montaña Blanca, Tenerife. *Bulletin of Volcanology* **57**, 337–55.
Ablay, G. J. & J. Marti 2000. Stratigraphy, structure and volcanic evolution of the Teide–Pico Viejo formation, Tenerife, Canary Islands. *Journal of Volcanology and Geothermal Research* **103**, 175–208.
Ancochea, E., J. L. Brandle, C. R. Cubas, F. Hernan, M. J. Huertas 1996. Volcanic complexes in the eastern ridge of the Canary Islands: the Miocene activity of the island of Fuerteventura. *Journal of Volcanology and Geothermal Research* **70**, 183–204.
Ancochea, E., J. M. Fuster, E. Ibarrola, A. Cendrero, J. Coello, F. Hernan, J. M.

Cantagrel, C. Jamond 1990. Volcanic evolution of the Island of Tenerife (Canary Islands) in the light of new K–Ar data. *Journal of Volcanology and Geothermal Research* **44**, 231–49.

Arana, V. & J. C. Carracedo 1978. *Canarian volcanoes I: Tenerife*. Madrid: Editorial Rueda.

Arana, V. & J. C. Carracedo 1978. *Canarian volcanoes II: Gran Canaria*. Madrid: Editorial Rueda.

Arana, V. & J. C. Carracedo 1979. *Canarian volcanoes III: Lanzarote and Fuerteventura*. Madrid: Editorial Rueda.

Blumenthal, M. M. 1961. Rasgos principales de la geología de las Islas Canarias con datos sobre Madeira. *Boletín Geológico y Minero* **72**, 1–30.

Bonelli Rubio, J. M. 1950. *Contribucion al estudio de la erupcion del Nambroque o San Juan (isla de La Palma)* ["Contribution to the study of the eruption of Nambroque or San Juan (island of La Palma)"]. Madrid: Instituto Geografico y Catastral.

Bravo, T. 1964. Estudio geológico y petrográfico de la isla de La Gomera. *Estudios Geológicos* **20**, 1–56.

Cantagrel, J. M., A. Cendrero, J. M. Fúster, E. Ibarrola, C. Jamond 1984. K–Ar chronology of the volcanic eruptions in the Canarian Archipelago: Island of La Gomera. *Bulletin Volcanologique* **47**, 597–609.

Cantagrel, J. M., N. O. Arnaud, E. Ancochea, J. M. Fuster, M. J. Huertas 1999. Repeated debris avalanches on Tenerife and genesis of Las Cañadas caldera wall (Canary Islands). *Geology* **27**, 739–42.

Carracedo, J. C. 1994. The Canary Islands: an example of structural control on the growth of large oceanic island volcanoes. *Journal of Volcanology and Geothermal Research* **60**, 225–41.

Carracedo, J. C. 1999. Growth, structure, instability and collapse of Canarian volcanoes and comparisons with Hawaiian volcanoes. *Journal of Volcanology and Geothermal Research* **94**, 1–19.

Carracedo, J. C. & E. Rodríguez Badiola 1991. *Lanzarote: la erupción volcánica de 1730* [with a colour 1:25 000 geological map of the eruption]. Madrid: Editorial MAE.

Carracedo, J. C., E. Rodríguez Badiola, V. Soler 1990. Aspectos volcanológicos y estructurales, evolución petrológica e implicaciones en riesgo volcánico de la erupción de 1730 en Lanzarote, Islas Canarias. *Estudios Geológicos* **46**(1–2), 25–55.

Carracedo, J. C., E. Rodríguez Badiola, V. Soler 1992. The 1730–36 eruption of Lanzarote, Canary Islands: a long, high-magnitude basaltic fissure eruption. *Journal of Volcanology and Geothermal Research* **53**, 239–50.

Carracedo, J. C., S. J. Day, H. Guillou, E. Rodríguez Badiola 1996. The 1677 eruption of La Palma, Canary Islands. *Estudios Geologicos* **52**, 103–114.

Carracedo, J. C., S. J. Day, H. Guillou, E. Rodríguez Badiola, J. A. Canas, F. J. Perez Torrado 1998. Hotspot volcanism close to a passive continental margin: the Canary Islands. *Geological Magazine* **135**, 591–604.

Carracedo, J. C., S. J. Day, H. Guillou, P. Gravestock 1999a. The later stages of the volcanic and structural evolution of La Palma, Canary Islands: the Cumbre Nueva giant collapse and the Cumbre Vieja Volcano. *Geological Society of America, Bulletin* **111**, 755–68.

Carracedo, J. C., S. J. Day, H. Guillou, F. J. Perez Torrado 1999b. Giant Quaternary landslides in the evolution of La Palma and El Hierro, Canary Islands. *Journal of Volcanology and Geothermal Research* **94**, 169–90.

Carracedo, J. C., E. Rodríguez Badiola, H. Guillou, J. de la Nuez, F. J. Perez Torrado 2001. Geology and volcanology of La Palma and El Hierro, Canary Islands. *Estudios Geologicos* **57**, 171–295. [Includes geological maps in colour.]

Carracedo, J. C., F. J. Pérez Torrado, E. Ancochea, J. Meco, F. Hernán, C. R. Cubas, R. Casillas, E. Rodríguez Badiola 2002. Cenozoic volcanism II: the Canary Islands. In *The geology of Spain*, F. A. W. Gibbons & T. Moreno (eds). London: Geological Society of London.

Cendrero, A. 1971. The volcano-plutonic complex of La Gomera (Canary Islands). *Bulletin Volcanologique* **34**, 537–61.

Cubas, C. R. 1978. Estudio de los domos sálicos de la Isla de La Gomera (Islas Canarias), I: vulcanología. *Estudios Geológicos* **34**, 53–70.

Damnati, B., N. Petit-Maire, M. Fontugne, J. Meco, D. Williamson 1996. Quaternary palaeoclimate in the eastern Canary Islands. *Quaternary International* **31**, 37–46.

Day, S. J., J. C. Carracedo, H. Guillou 1997. Age and geometry of an aborted rift flank collapse: the San Andres fault, El Hierro, Canary Islands. *Geological Magazine* **134**, 523–37.

Day, S. J., J. C. Carracedo, H. Guillou, P. Gravestock 1999. Recent structural evolution of the Cumbre Vieja volcano, La Palma, Canary Islands: volcanic rift zone reconfiguration as a precursor to volcano flank instability? *Journal of Volcanology and Geothermal Research* **94**, 135–67.

Féraud, G., G. Giannerini, R. Campredon, C. J. Stillman 1985. Geochronology of some Canarian dyke swarms: contribution to the volcano-tectonic evolution of the archipelago. *Journal of Volcanology and Geothermal Research* **25**, 29–52.

Guillou, H., J. C. Carracedo, F. J. Perez Torrado, E. Rodríguez Badiola 1996. K–Ar ages and magnetic stratigraphy of a hotspot – induced, fast grown oceanic island: El Hierro, Canary Islands. *Journal of Volcanology and Geothermal Research* **73**, 141–55.

Guillou, H., J. C. Carracedo, R. Duncan 2001. K–Ar, $^{40}Ar/^{39}Ar$ Ages and magnetostratigraphy of Brunhes and Matuyama lava sequences from La Palma Island. *Journal of Volcanology and Geothermal Research* **106**, 175–94.

Hausen, H. 1971. Outlines of the geology of Gomera. *Societas Scientiarum Fennica, Commentationes Physico-Mathematicae* **41**, 1–53.

Hoernle, K. A. & H–U. Schmincke, 1993. The role of partial melting in the 15 Ma geochemical evolution of Gran Canaria: a blob model for the Canary hotspot. *Journal of Petrology* **34**, 599–626.

Klügel, A., H–U. Schmincke, J. D. L. White, K. A. Hoernle 1999. Chronology and volcanology of the 1949 multivent rift zone eruption on La Palma, Canary Islands. *Journal of Volcanology and Geothermal Research* **94**, 267–82.

Krastel, S., H–U. Schmincke, C. L. Jacobs, R. Rihm, T. P. Le Bas, B. Alibes 2001. Submarine landslides around the Canary Islands. *Journal of Geophysical Research* **106**, 3977–97.

Marti, J., J. Mitjavila, V. Arana 1994. Stratigraphy, structure and geochronology of the Las Cañadas caldera (Tenerife, Canary Islands). *Geological Magazine* **131**, 715–27.

McDougall, I. & H–U. Schmincke 1976. Geochronology of Gran Canaria, Canary Islands: age of shield building volcanism and other magmatic phases. *Bulletin*

Volcanologique **40**, 57–77.

McGuire, W. J. 1996. Volcano instability: a review of contemporary themes. In *Volcano instability on the Earth and other planets*, W. J. McGuire, A. P. Jones, J. Neuberg (eds), 1–23. Special Publication 110, Geological Society, London.

Meco, J., H. Guillou, J. C. Carracedo, A. Lomoschitz, A. J. G. Ramos, J. J. Rodríguez-Yánez 2002. The maximum warmings of the Pleistocene world climate recorded in the Canary Islands. *Palaeogeography, Palaeoclimatology, Palaeoecology* **2885**, 1–14.

Mehl, K. W. & H-U. Schmincke 1999. Structure and emplacement of the Pliocene Roque Nublo debris avalanche deposit, Gran Canaria, Spain. *Journal of Volcanology and Geothermal Research* **94**, 105–134.

Moore, J. G. 1964. Giant submarine landslides in the Hawaiian ridge. Professional Paper 501-D (pp. 95–98), US Geological Survey, Denver, Colorado.

Moore, J. G., W. R. Normark, R. T. Holcomb 1994. Giant Hawaiian landslides. *Annual Reviews of Earth and Planetary Science* **22**, 119–44.

Perez Torrado, F. J., J. C. Carracedo, J. Mangas 1995. Geochronology and stratigraphy of the Roque Nublo Cycle, Gran Canaria, Canary Islands. *Geological Society of London, Journal* **152**, 807–818.

Perez Torrado, F. J., J. Marti, J. Mangas, S. J. Day 1997. Ignimbrites of the Roque Nublo Group, Gran Canaria, Canary Islands. *Bulletin of Volcanology* **58**, 647–54.

Scarth, A. 1999. *Vulcan's fury: man against the volcano*. London: Yale University Press.

Scarth, A. & J-C. Tanguy 2001. *Volcanoes of Europe*. Harpenden, England: Terra.

Schirnick, C., P. van den Bogaard, H-U. Schmincke 1999. Cone sheet formation and intrusive growth of an oceanic island: the Miocene Tejeda complex on Gran Canaria (Canary Islands). *Geology* **27**, 207–210.

Schmincke, H-U. 1982. Volcanic and chemical evolution of the Canary Islands. In *Geology of the northwest African margin*, U. von Rad, K. Hinz, M. Sarnthein, E. Seibold (eds), 273–306. Berlin: Springer.

Staudigel, H. & H-U. Schmincke 1984. The Pliocene Seamount Series of La Palma, Canary Islands. *Journal of Geophysical Research* **89**, 11195–215.

Steiner, C., A. Hebsen, P. Favre, G. M. Stampfli, J. Hernández 1998. Mesozoic sequence of Fuerteventura: witness of Early Jurassic sea floor spreading in the central Atlantic. *Bulletin of the Geological Society of America* **110**, 1304–317.

Stillman, C. J. 1987. A Canary Island dyke swarm: implications for the formation of oceanic islands by extensional fissural volcanism. In *Mafic dyke swarms*, H. C. Halls & W. F. Fahrig (eds), 243–55. Special Paper 34, Geological Association of Canada, Toronto.

Stillman, C. J. 1999. Giant Miocene landslides and the evolution of Fuerteventura, Canary Islands. *Journal of Volcanology and Geothermal Research* **94**, 89–104.

Urgeles, R., M. Canals, J. Baraza, B. Alonso, D. G. Masson 1999. The most recent mega-landslides of the Canary Islands: El Golfo debris avalanche and Canary debris flow, west El Hierro Island. *Journal of Geophysical Research* **102**, 20305–323.

Watts, A. B. & D. G. Masson 1995. A giant landslide on the north flank of Tenerife, Canary Islands. *Journal of Geophysical Research* **100**, 24487–98.

Index